# Pollution and Control

A social history of the Thames
in the nineteenth century

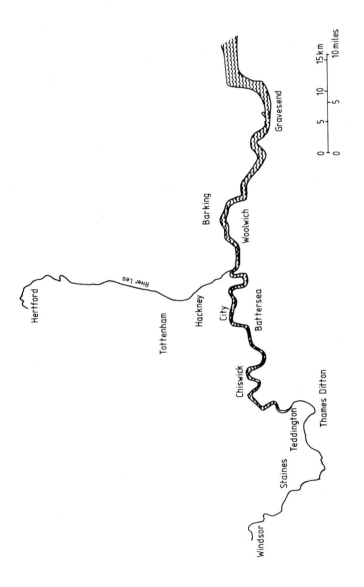

The Thames and the Lea, showing the general areas discussed in the text.

# Pollution and Control

A social history of the Thames
in the nineteenth century

## Bill Luckin

Adam Hilger, Bristol and Boston

*British Library Cataloguing in Publication Data*

Luckin, Bill
    Pollution and control: a social history of the Thames in the nineteenth century.
    1. Water——Pollution——England——Thames, River
    ——History——19th century   2. Water——Pollution
    ——England——London——History——19th century
    I. Title
    628.1′688′421   TD264.TS
    ISBN 0-85274-472-2

Consultant Editor: **Professor A J Meadows,**
                University of Leicester

Published under the Adam Hilger imprint by IOP Publishing Limited
Techno House, Redcliffe Way, Bristol BS1 6NX, England
PO Box 230, Accord, MA 02018, USA

Typeset by BC Typesetting
51 School Road, Oldland Common, Bristol BS15 6PJ

Printed in Great Britain by J W Arrowsmith Ltd, Bristol

For Lesley, David and Steven

# Contents

# Preface and Acknowledgments

Pollution and purity; deterioration and regeneration; order and disorder; health and disease—it is around these enduring dichotomies and their social, political and scientific formations that this study of the Thames in the nineteenth century is organised. The preparation and writing of it have involved an attempt to link social and political history, historical demography and the social history of disease. If I have failed to make valid connections the fault is mine and emphatically not to be attributed to those who have talked with me about history during the last decade and a half. At Cambridge in the mid-1960s Joseph Needham taught me far more than he realised and later, in the same city, my friends Peter Searby and Tom Dunne offered invaluable encouragement. John Krige, Russell Moseley and Brian Easlea introduced me to the complexities of the social history of science at the University of Sussex: and, latterly, in the north west, John Pickstone, Roger Cooter and Stella Butler have provided much valued stimulus and friendship. I would also like to thank the organisers of the research seminars who have invited me to give papers based, directly or indirectly, on my work on the Thames, and my full and part-time students at Bolton who have made me think hard about how to write, rather than simply describe, interdisciplinary history. I am also grateful to the Research and Development Committee at Bolton for granting me a term's leave to complete the manuscript. The final draft was expertly typed by Margaret Cooter. To the late H J Dyos, I owe a great deal, since it was he who first persuaded me that a book of this type was worth writing and that I could and should write it. Grateful thanks are due, finally, to David Reeder for an astute assessment both of the content and structure of the manuscript.

This study could not have been completed without the unstinting assistance of officers and staff at the following libraries and institutions: the GLC History Library; the Thames Conservancy Archive, Wapping; the Society of Community Medicine; the University Library, Cambridge; the British Library; and the Public Record Office at Kew. Parts of Chapter 4 originally appeared in a slightly different form in 'The Final Catastrophe—Cholera in London, 1866', *Medical History* **21** (1977) 32–42, and are reproduced by kind permission of the editors of that journal. Unless otherwise stated, all publications cited in the text and bibliography are published in London.

**Bill Luckin**

# Introduction

It is now more than 50 years since John and Barbara Hammond, in the course of their debate with Sir John Clapham over the impact of industrialisation on standards of living in Britain during the nineteenth century, conceded the 'quantitative' case, but insisted that the 'qualitative' should remain open.[1] When, during the late 1950s, the controversy was reactivated in the form of an interchange of articles between R M Hartwell and E J Hobsbawm, it was still the statistical and the overtly measurable—changing levels of population and national and *per capita* income—which predominated.[2] It was only a decade or so afterwards that it came to be at all widely recognised that other, less easily quantifiable variables, namely levels of mortality and morbidity and environmental change and deterioration within the new urban–industrial society, demanded further, detailed research.[3] Why British economic and social historians should have given so little attention to fundamental interactions between men and women and the environment, as well as to the social and political structures and pressures which partly determine such interactions, is a complicated question; and one which becomes more rather than less puzzling when it is remembered that during the late 1960s and the 1970s there was an explosion of writing in the social sciences concerned with worldwide ecological and energy problems.[4] In general, however, British researchers, a minority of them strongly under the influence of American 'new economic historians', remained preoccupied with explicitly quantitative rather than environmental or qualitative aspects of the past.[5]

French historiography followed a different and distinctive course. Only a minority of scholars working in economic and social history in that country have until recently devoted themselves to statistical and causal aspects of industrialisation.[6] This may be explained partly in terms of the actual history of France and her failure to undergo an industrial revolution in the late eighteenth and early nineteenth

1

centuries but it is related also to the eclecticism of a dominant and generally non-positivistic historical methodology and the unforced and genuinely interdisciplinary relationship between history and the other human sciences. The magisterial work of Fernand Braudel, with its emphasis on the interconnectedness of human, geological and geographical structures[7] and of Emmanuel Le Roy Ladurie, with its exploration of the impact of climatic and autonomous epidemiological transformations on peasant economy and society, is pre-eminent.[8] (Highly important, also, within a different but no less 'European' tradition, are the writings of Jerome Blum, which give a superb account of the relationship between men, women and brute nature under the Old Regime in eastern and western Europe.[9])

If British scholarship cannot boast an impressive body of literature in environmental history, different conclusions may be arrived at in relation to the subdiscipline conventionally and not altogether happily known as the 'history of public health'. Here biographical and institutional studies by R A Lewis[10], S E Finer[11], the late Royston Lambert[12] and John Eyler[13] have been complemented by general surveys by George Rosen[14], F B Smith[15] and A S Wohl.[16] What is especially impressive about Smith's and Wohl's work is that it has drawn upon recent findings in urban, demographic and sociomedical history to provide insights into disease and medical intervention during the peak period of industrialisation which transcend the often Whiggish conclusions of conventional histories of public health. (Wohl's book, incidentally, is the first of its kind to contain chapters on the pollution of air and water.[17])

There are clear connections between environmental history, the history of disease and historical demography: and the last of these has been one of the major growth areas in British, French, American and German social history during the last 30 years.[18] Although its findings are still the subject of wide-ranging debate, the massive 'reconstruction' of the English population between 1541 and 1871 by E A Wrigley and R S Schofield is now recognised as a *locus classicus*.[19] In terms, however, of aggregate and cause-specific mortality and the environmental determinants of the reduction in deaths from infectious disease during the nineteenth century, the work of Thomas McKeown is central[20]: and latterly, though belatedly, a small number of historians have begun to test McKeown's over-arching hypotheses against national and regional experience.[21]

Others, however (and here there has perhaps been an undue emphasis on cholera), have been preoccupied less with epidemiological and environmental processes than with the social and political meanings and ramifications of decimating infection in the past. A now thriving subdiscipline in the social history of medicine has concentrated *inter alia* on the extent to which notions and images of disease are constructed or shaped by processes external to institutionalised medicine—the relationship between dominant class formations and ideas about the transmission of infection, and interactions between medical and other bodies of human thought and culture. Its intellectual pedigree is complex but Marxism[22], the sociology of knowledge as developed by Karl Mannheim[23], and Thomas Kuhn's work on the nature of scientific revolutions and normal science have all been formative.[24] So, also, have the writings of the late Michel Foucault who, more than any other thinker of his generation, has forced social historians of science to ponder the relationship between specialist knowledge, the language in which it is expressed, and the ambiguities of reform.[25] If social history and the sociology of medicine are relevant to the central theme of this book so, equally, is social anthropology. The connections between social history and social anthropology, notably in the writings of Keith Thomas[26] and Alan Macfarlane[27], are now well established in this country. But, in relation to pollution and, more specifically, the question 'Why is it that *this* aspect of the external world is deemed polluting at *this* particular historical moment?' the work of Mary Douglas, with its emphases on order and disorder, interactions between the 'natural' and the 'social', and the mediations between social structure and taboo, is seminal.[28] (Why historical writing on the perception of the environment and, by extension, on environmentalism, should still be so under-developed is itself a fascinating question in the sociology of knowledge.[29])

What of the relations between environmental and urban history? It is now generally acknowledged that urban history in this country was greatly stimulated by that highly active and open-minded scholar, the late H J Dyos.[30] Although explanations cast exclusively in terms of individuals or academic institutions are rarely convincing there can be little doubt that Dyos' Urban History Group opened the eyes of a generation of historians to the possibility of collaborating with disciplines—social and historical geography, urban sociology and anthropology, town planning—from which they had been too long insulated, and of concentrating on aspects of urban–industrial society, notably the spatial, the visual and the occupational, which were still underdeveloped.[31] At a time when all forms of academic history are under siege, with the loss of tenured posts in social history and historical geography biting deep into the quality of our culture, it should be remembered that Dyos, who was a liberal in the genuinely humane sense of that complex word, welcomed the writings of two of the most influential Marxist historians of our times, Gareth Stedman Jones[32] and John Foster[33].

It is necessary, also, to comment on the historiography of London during the nineteenth century and on the problems of writing a social history of a great river like the Thames. It has become a truism to say that, compared with the towns and cities of the English North and the Midlands, the capital has been ill-served by historians. Certainly, with the exception of Stedman Jones' fine work on *Outcast London*, there are no metropolitan equivalents to Armstrong's York[34], Anderson's Preston[35], or Hennock's Birmingham.[36] The underlying reasons are probably that the sheer size of the capital has daunted attempts at comprehensive history-writing and that its institutional complexity— much more complex than even the most labyrinthine and committee-bound municipality in the nineteenth century—has been an added disincentive. Having said that, one must immediately draw attention to the writings of Francis Sheppard[37] and to David Owen's posthumous study of the Metropolitan Board of Works.[38]

As for the Thames, the bibliography at the end of this book reveals a venerable and fascinating literature, but one which is heavily weighted towards anecdotes, boats and riverside rambles. My original intention was to communicate something of the capital and its social structure through an analysis of the recurring crises precipitated by its major river. The only approximate model that came to mind was Richard Cobb's vivid evocation of the Seine in the late eighteenth century and this convinced me that an explicitly social and environmental history of a river could indeed be written.[39] Whether the perceptions of the Thames presented here do throw light on the institutions and distinctive class formations of London during the nineteenth century will only be decided by the individual reader.

A final introductory issue requires clarification. Edwin Chadwick, often acknowledged as central to the development of proto-environ-mentalism, looms small in this book. This can be explained partly in terms of subject matter and method—this is a study concerned more with the condition of the Thames during the nineteenth century as a whole than with Chadwick's public health 'dictatorship' during the late 1840s and earlier 1850s. But a playing down of his role can also be justified in terms of the history and dissemination of ideas. Relatively few public health workers (medical officers, doctors, scientists) described and quoted in what follows appear to have been heavily influenced by Chadwick's total urban–environmental 'system'. This raises the question of the extent to which his idealised 'circular' scheme—a constant supply of piped water, water-carried sewage and the 'enrichment' of agricultural land with the resultant human waste— can be said to have possessed genuine originality. His 'grand strategy' for London may have been nothing more or less than a canny and, in some respects, commonsensical yoking together of generalised medical and environmental doctrines long current among those with an interest in public health: doctrines which would retain their potency, independently of Chadwick's name and charisma, into the 1870s and beyond.[40]

This study is organised into three parts. The first consists of a survey of the problem of the pollution of the Thames between about 1830 and 1900. An introductory survey is followed by two chapters—on water filtration and water analysis—which seek to identify that problem at a higher level of specificity. The second part is concerned with demographic experience, and, more particularly, with those water-transmitted and water-related infections, cholera, diarrhoea and typhoid, the incidence of which was determined, to a greater or lesser extent, by the state of the Thames. The discussion of the cholera epidemic of 1866 contains a lengthy analysis of the state of the Lea—the justification here is, firstly, that no account of epidemics triggered by unsafe water in London during the nineteenth century can afford to omit this outbreak and, secondly, that the numerous inquiries into the causes of the crisis had a significant effect both on attitudes towards the Thames and the 'water question' in general. The third and concluding part deals with the 'politics of control'. The first chapter in this part examines the prevention of pollution of the Thames during the second half of the nineteenth century, and, at a more detailed level, the protracted and at times bizarre conflict between the Thames Conservancy Board and the Metropolitan Board of Works; the second explains why it was that national legislative provision for the protection of rivers in nineteenth-century Britain was so extraordinarily weak.

1   J L Hammond, 'The Industrial Revolution and Discontent', *Econ. Hist. Rev.* **2**(2) (1930) 215–28.

2   The debate is now easily accessible in E J Hobsbawm, *Labouring Men* (1968) chapters 5–7 and R M Hartwell, *The Industrial Revolution and Economic Growth* (1971). A useful overview is provided in A J Taylor (ed), *The Standard of Living in Britain in the Industrial Revolution* (1975).

3   See the perceptive comments of G N von Tunzelmann, 'Trends in Real Wages 1750–1850, Revisited', *Econ. Hist. Rev.* **32** (1979) 49.

4   The 'limits to growth' debate was central here. For relevant bibliographies see Paul R Erlich and Anne H Erlich, *Population, Resources, Environment: Issues in Human Ecology* (San Francisco, 1970).

5   But one debate in the 'new economic history'—precipitated by the publication of Robert William Fogel and Stanley L Engerman, *Time on the Cross: The Economics of American Negro Slavery* (1974)—generated reflection on and research into some 'qualitative' and environmental areas.

6    Note, however, the contribution to British economic and social history of François Crouzet and especially his 'England and France in the Eighteenth Century: A Comparative Analysis of Two Economic Growths' in R M Hartwell (ed), *The Causes of the Industrial Revolution in England* (1967). See also F Crouzet (ed), *Capital Formation in the Industrial Revolution* (1972) and F Crouzet, *The First Industrialists: The Problem of Origins* (Cambridge, 1985).

7    The *locus classicus* is *The Mediterranean and the Mediterranean World in the Age of Philip II* vols I and II (translated by S Reynolds, 1972). But see also the synoptic *Capitalism and Material Life 1400–1800* (translated by Miriam Kochan, 1973). On methodology there is F Braudel, *Afterthoughts on Material Civilization and Capitalism* (translated by P M Ranum, Baltimore, 1977).

8    Ladurie is formidably productive. The 'environmental' aspects of his approach are especially evident in two collections of essays, *The Territory of the Historian* (translated by S and B Reynolds, 1979) and *The Mind and Method of the Historian* (translated by S and B Reynolds, 1981). But see also *Times of Feast, Times of Famine: A History of Climate since the Year 1000* (translated by B Bray, 1972).

9    Jerome Blum, *The End of the Old Order in Rural Europe* (Princeton, 1978) and especially Part I.

10   R A Lewis, *Edwin Chadwick and the Public Health Movement 1832–1854* (1952).

11   S E Finer, *The Life and Times of Sir Edwin Chadwick* (1952).

12   Royston Lambert, *Sir John Simon 1816–1904 and English Social Administration* (1963).

13   J M Eyler, *Victorian Social Medicine: The Ideas and Methods of William Farr* (Baltimore, 1979).

14   George Rosen, *A History of Public Health* (New York, 1958) and 'Disease, Debility and Death' in H J Dyos and Michael Wolff (eds), *The Victorian City: Images and Reality* vol II (1973) 625–67.

15   F B Smith, *The People's Health 1830–1910* (1979).

16   Anthony S Wohl, *Endangered Lives: Public Health in Victorian Britain* (1983).

17   *Ibid*, chapters 8 and 9.

18   The late M W Flinn cites over 700 items in *The European Demographic System 1500–1820* (1981).

19   E A Wrigley and R S Schofield, *The Population History of England 1541–1871: A Reconstruction* (1981). But see 'The Population History of England 1541–1871: A Review Symposium', *Social History* **8** (1983) 139–68 and M W Flinn, 'The Population History of England, 1541–1871', *Econ. Hist. Rev.* **35** (1982) 443–57.

20   Thomas McKeown, *The Modern Rise of Population* (1976).

21   Seminal here is Peter Razzell, *The Conquest of Smallpox* (Firle, 1977). This should be read in conjunction with the same author's *Edward Jenner's Cowpox Vaccine: The History of a Medical Myth* (Firle, 1977). See also R Woods and J Woodward (eds), *Urban Disease and Mortality in Nineteenth Century England* (1984). A new set of transnational hypotheses have

recently been developed by William H McNeill in his *Plagues and Peoples* (Oxford, 1977) but these will be exceptionally difficult to test.

22   An important contributor here has been Robert M Young. See, in particular, his 'The Historiographic and Ideological Contexts of the Nineteenth Century Debate on Man's Place in Nature' in M Teich and R M Young (eds), *Changing Perspectives in the History of Science* (1973) 344–438 and 'Science *is* Social Relations', *Radical Sci. J.* **5** (1977) 65–129. An excellent overview of radical writing in the sociology and social history of medicine is provided in Karl Figlio, 'Sinister Medicine? A Critique of Left Approaches to Medicine', *Radical Sci. J.* **9** (1979) 14–68. The same author's 'Chlorosis and Chronic Disease in Nineteenth Century Britain: The Social Constitution of Somatic Illness in a Capitalist Society', *Social History* **3** (1978) 167–97, is a closely argued account of the interaction between dominant class structures and disease formations. See also Roger Cooter, *The Cultural Meaning of Popular Science: Phrenology and the Organization of Consent in Nineteenth Century Britain* (Cambridge, 1984).

23   Karl Mannheim, *Ideology and Utopia* (translated by Louis Worth and Edward Shils, 1936). A recent monograph by A P Simonds, *Karl Mannheim's Sociology of Knowledge* (Oxford, 1978), elucidates the strengths and weaknesses of Mannheim's corpus.

24   Thomas S Kuhn, *The Structure of Scientific Revolutions* (Chicago, 1962). Within the post-Kuhnian tradition see, in particular, S B Barnes, *Scientific Knowledge and Sociological Theory* (1974), *Interests and the Growth of Knowledge* (1977) and *T. S. Kuhn and Social Science* (1982). David Bloor, *Knowledge and Social Imagery* (1976) is particularly relevant to the first section of the present study. A perceptive discussion of recent developments in the field as a whole is provided by Michael Mulkay in *Science and the Sociology of Knowledge* (1979).

25   In the present context one should note Michel Foucault, *Madness and Civilization: A History of Insanity in the Age of Reason* (translated by Richard Howard, 1967), *The Order of Things: The Archaeology of the Human Sciences* (1970) and *The Birth of the Clinic: An Archaeology of Medical Perception* (translated by A M Sheridan, 1973).

26   Keith Thomas, *Religion and the Decline of Magic: Studies in Popular Beliefs in Sixteenth- And Seventeenth-Century England* (1971) and *Man and the Natural World: Changing Attitudes in England 1500–1800* (1983).

27   Alan Macfarlane, *Witchcraft in Tudor and Stuart England* (1970). The same author's *The Origins of English Individualism* (Oxford, 1978) is controversial but its opening chapter raises numerous problems generated by the cross-fertilisation between history and anthropology. James Obelkevich, *Religion and Rural Society: South Lindsey 1825–1875* (Oxford, 1976) is an accomplished monograph devoted to a later period which makes telling use of a broadly anthropological mode.

28   See, in particular, Mary Douglas, *Purity and Danger: An Analysis of Concepts of Pollution and Taboo* (1966), *Implicit Meanings: Essays in Anthropology* (1975) and Mary Douglas and Aaron Wildavsky, *Risk and Culture:*

*An Essay on the Selection of Technological and Environmental Dangers* (Berkeley, 1982).

29    But see L J Jordanova and Roy S Porter (eds), *Images of the Earth: Essays in the History of the Environmental Sciences* (BSHS Monographs **I** (1979)), John Passmore, *Man's Responsibility for Nature* (1974) and Keith Thomas (1983) (note 26 above). On the history of administering the environment see Roy M MacLeod, 'The Alkali Acts Administration 1863–84: The Emergence of the Civil Scientist', *Victorian Stud.* **9** (1965) 85–112 and 'Government and Resource Conservation: The Salmon Acts Administration' *J. Br. Stud.* **7** (1968) 144–50.

30    His approach is demonstrated in the posthumous collection of essays, David Cannadine and David Reeder (eds), *Exploring the Urban Past: Essays in Urban History* (Cambridge, 1982).

31    But here one must note the pioneering work of Francis D Klingender in *Art and the Industrial Revolution* (1947).

32    Gareth Stedman Jones, *Outcast London: A Study in the Relationship between Classes in Victorian Society* (Oxford, 1971). See also the same author's *Languages of Class: Studies in English Working Class History 1832–1982* (Cambridge, 1983). Much of what is said about 'rhetorics' in Chapter 7 of the present study has been influenced by Stedman Jones' important essay, 'Rethinking Chartism', in that volume.

33    John Foster, *Class Struggle and the Industrial Revolution: Early Industrial Capitalism in Three English Towns* (1974).

34    Alan Armstrong, *Stability and Change in an English Town: A Social Study of York 1801–1851* (Cambridge, 1974). This book is one of the very few of its kind to provide a comprehensive account of patterns of mortality within the new urban environment.

35    Michael Anderson, *Family Structure in Nineteenth Century Lancashire* (Cambridge, 1971).

36    E P Hennock, *Fit and Proper Persons: Ideal and Reality in Nineteenth Century Urban Government* (1973).

37    F W H Sheppard, *Local Government in St Marylebone 1688–1835, A Study of the Vestry and the Turnpike Trusts* (1958) and Francis Sheppard, *London 1808–1870; The Infernal Wen* (1971).

38    David Owen, *The Government of Victorian London 1855–1889: The Metropolitan Board of Works, the Vestries and the City Corporation* (edited by Roy MacLeod, Cambridge, MA, 1982).

39    Richard Cobb, 'The Seine' in *Paris and its Provinces 1792–1802* (Oxford, 1975) 57–86.

40    The best account of Chadwick as 'environmentalist' is to be found in R A Lewis (note 10). But there is considerable contextual value in Margaret Pelling, *Cholera, Fever and English Medicine 1825–1865* (Oxford, 1978), chapters 1 and 2 and Graeme Davison, 'The City as a Natural System', in Derek Fraser and Anthony Sutcliffe (eds), *The Pursuit of Urban History* (1983) and particularly p 364 *passim*.

# Part I

# Crisis and After

If we can abstract pathogenicity and hygiene from our notion of dirt, we are left with the old definition of dirt as matter out of place. This is a very suggestive approach. It implies two conditions: a set of ordered relations and a contravention of that order. Dirt, then, is never a unique, isolated event. Where there is dirt there is system.

Mary Douglas, *Purity and Danger: An Analysis of Concepts of Pollution and Taboo* (1966) 35.

# 1 The State of the River

This chapter surveys the history of the pollution of the Thames between the early nineteenth and the beginning of the twentieth century. It is concerned as much with how the specific social problem came to be constituted and linked to social and political processes as with the condition of the river between and within decades. It might have been possible to have constructed an index to show changing levels of contamination during this period but such an exercise would have been less revealing than an account of why some contemporaries believed the river to be irredeemably damaged while others insisted that it could be reclaimed and 'repurified'.

The first part of this chapter deals with the period which began with widespread anxiety over the state of the Thames in the 1820s and ended with the crisis of 1858 when Parliament was forced finally to take action against the foulness of the inner-city river; attention is also given here to the ways in which images of the natural world were appropriated and deployed within the political domain. This is followed by an analysis of the 1860s and 1870s, when perceptions of pollution in London underwent modification, and medical men and metropolitan medical officers of health were divided into those who took up broadly optimistic and pessimistic attitudes towards the 'water question'. The theme of 'domestic pollution'—whereby supplies delivered in a moderately safe condition allegedly underwent deterioration in unsatisfactory storage receptacles in individual homes—is also discussed and linked, in a final subsection, to an examination of the body of scientific ideas which provided the framework, during the final 25 years of the century, for a continuing debate about the healthiness or otherwise of the Thames and the drinking water taken from it. Here, again, the optimistic–pessimistic dichotomy, as well as the relative credibility of claims based on bacteriological, statistical or epidemiological theory, continued to be dominant.

11

By the end of the first quarter of the nineteenth century there was strong
evidence, even when allowance is made for exaggerated claims for the
existence of a wholly pure and pristine river during the eighteenth
century, that the Thames had become unprecedentedly polluted.
An especially damning indictment was that there had been 'an entire
destruction of the fisherman's trade between Putney Bridge and
Greenwich'.[1] At the beginning of the century, or so it was claimed by
John Goldham, clerk to Billingsgate Market, the salmon industry had
flourished, with fish selling at between three or four shillings a pound.
Sizable numbers of plaice, dace and dabs had also been caught.
As many as 400 fishermen had worked the river: apprenticeship had
thrived and a single boat had been able to earn up to £6 a week. By the
late 1820s this had been reduced by half and boats and nets were
already being sold off.[2] Even that most hardy of river creatures, the eel,
was reported to be perishing from lack of oxygen.[3] And so suspicious
had several of the great metropolitan brewers become of Thames water
that they had been forced, 'at vast expense', to look to wells for their
supplies.[4]

**Table 1.1** The population of London 1800–1911. Data from
B R Mitchell *The Fontana Economic History of Europe* vol 4 Statistical
Appendix (1971), 17 and *Annual Reports* and *Supplements* of the
Registrar-General.

| Year | 'Greater' London | Year | 'Registration' London |
|------|------------------|------|------------------------|
| 1800 | 1117 000 | 1851 | 2362 236 |
| 1850 | 2685 000 | 1861 | 2583 112 |
| 1880 | 4770 000 | 1871 | 3254 260 |
| 1910 | 7256 000 | 1881 | 3816 483 |
|      |          | 1891 | 4211 056 |
|      |          | 1901 | 4536 429 |
|      |          | 1911 | 4521 685 |

When they searched for explanations of this deterioration contemporaries probably exaggerated the impact of industrial effluent. There
could be no doubt that the gas and steamboat companies had added
greatly to the filth of the river since the early nineteenth century[5], or
that 'the refuse of hospitals, slaughterhouses, colour, lead and soap-
works and manufacturies' had also played their part.[6] But it was the
sheer pressure of population and associated volumes of sewage, flowing out into the river by way of ever larger numbers of water-closets
and uncontrolled sewers that had wrought the most severe damage
(table 1.1). Looking back from the vantage point of the early 1840s,

Thomas Cubitt was nostalgic for the well-tried individual cesspool system which had protected the river from an excess of sewage pollution. Now, he claimed, 'the Thames is made a great cesspool instead of each person having one of his own'.[7]

By the late 1820s the inner-city Thames had become horribly turgid. 'Scarcely a week passes', a witness told the Royal Commission of 1828, 'but the carcass of one or more dead dogs is deposited within a short distance of my residence.' He had frequently seen 'the same carcass float up and down with the tide for several days, sometimes deposited a little above, then a little below, and this I have seen for 10 or 12 days.'[8] But in the absence of an agreed body of theory that the consumption of such water might lead to massive mortality, the private companies remained unapologetic and sometimes even lavished praise on the state of the river. 'The impregnating ingredients of the Thames', a medical spokesman for the Grand Junction Company insisted, 'are as perfectly harmless as any spring-water of the purest kind used in common life: indeed, there is probably not a spring, with the exception of Malvern, and one or two more, which are so pure as Thames water.'[9] Controversy over the degree of contamination of the river was, and would continue to be, bitter, with those who supported the companies and those who campaigned against their extensive and oligopolistic powers often seeming to refer to two wholly different versions of the physical world. It was, in fact, the economic and political hegemony of the water companies, combined with uncertainty as to how, precisely, unsafe water might adversely affect individual and communal health, which provides a general explanation of why the 'water question' was not more extensively investigated during the 1830s.[10] There was an unsuccessful attempt to establish a 'Metropolis Pure Soft Water Company' in 1834–5, and complaints both about the quality and quantity of supplies delivered by the New River Company were heard before the Common Council of the City of London in 1836.[11] In 1837 and 1838 deputations on similar topics, including the 'purification' of the Thames, were received by the Chancellor of the Exchequer.[12] But, taking the decade as a whole, pollution seems to have been of minor concern compared with campaigns in the fields of poor relief and the prevention of fever.[13]

But this may be to misread the form in which the problem was actually predominantly perceived in the period between the Royal Commission of 1828 and the large-scale urban surveys of the early 1840s. For what did not diminish at this time was widespread anxiety over one of the most pressing and intractable dilemmas of early industrial society—how to move vast volumes of human waste out and away from overcrowded streets, courts and alleys. The dominant ideology demanded that this was a task which should be undertaken by private companies sanctioned by Act of Parliament. Economic orthodoxy dictated that there must be a profitable interaction between the urban and the rural sectors, with the townsman selling his 'excess' sewage to the countryman so that it could be used as agricultural fertiliser. And aesthetic convention insisted that the profitable disposal of urban waste could be combined with projects which would both 'beautify' and 'purify' a potentially polluted and disorderly urban civilisation. Medical opinion might still be ambivalent as to the precise impact of dirty water on human health but river pollution was now increasingly seen as socially pernicious and perverse. The most acceptable scheme, therefore, would be that which would simultaneously rid the city of its potentially profit-laden effluent and 'repurify' both air and water.

Each of these aims was combined in a series of projects devised between 1827 and 1847 by John Martin, visionary and early Victorian artist of the 'sublime'.[14] It would be easy to consign Martin's ideas to the menagerie of the bizarre and the unbalanced, had they not integrated, in a vivid form, the economic, environmental and ideological components which have been outlined above, and revealed, albeit obliquely, pervasive perceptions of and attitudes towards the pollution of the Thames in the 1830s. Martin's plan, as presented to a select committee in 1834, was based on the construction of an intercepting sewage system which would be linked to two 'grand receptacles' on the Regent's and Grand Surrey Canals. From these two points, the city's waste would be transported in its entirety to the countryside and sold, at a profit, to farmers.[15] As each successive version of what one critic termed a 'Babylonish' conception[16] failed to gain economic support and parliamentary backing, Martin added terraces, bridges, weirs and promenades to his original blueprint.[17] Much that he said on his own behalf would no doubt have been better left unsaid; and, latterly, his thinking was heavily impeded by obsessiveness and paranoia. But there was an undeniably 'visionary' quality to his championing of an intercepting sewage system: and, had he excluded the chimera of agricultural disposal from his scheme, there would have been a marked similarity between his 'grand receptacles' and the outlets which would be built at Crossness and Barking in the 1860s. It should also be noted that commitment to 'beautification' and the profitability of urban waste continued, albeit in a less grandiose style, to influence the thinking of those responsible for the protection of health and environment in London during the next 40 years—there were tenders for contracts to dispose of at least a portion of London's sewage in 1863, 1867, 1872 and 1880.[18] When Martin had already despaired of seeing the project which he had first devised in 1827 being translated into reality he reflected on the way in which the Thames fitted into his scheme for the salvation of the early-nineteenth-century city and of the significance which he attached, in terms of ill-health and wasted resources, to polluted water. 'Does it not show incredible ignorance', he asked, 'that a country boasting itself second to none should not merely cast away its real wealth, namely the means of producing that which is actually useful and necessary, but that, in doing so, it should cause additional waste, by rendering a great part of the metropolis water unfit for use?'[19]

'Unfit for use'—certainly by 1840 there were those, lay and medical men alike, who were convinced that the Thames was profoundly and irrevocably polluted. The river was now also more widely seen as being detrimental to good health; so abject a state of filthiness and 'unnaturalness' reflected and itself further contributed to a loss of order and stability within the new urban civilisation. 'The Water that is coming in from the Thames', a Marylebone builder warned in 1840, 'wherein all the Privies, Water-closets and Drains from all the Houses go into the Thames, that cannot be such Water that the People of London ought to drink and drink again and again.'[20] 'Do you think the poor inhabitants of London are prevented from drinking the water supplied to them from finding objectionable matter in it?' an eminent authority was asked four years later. 'Certainly', came the reply.[21] Even worse, there were those, among the most wretched of the metropolitan poor, who lived close to 'some very filthy places near the river . . . among an immense quantity of mud and filth', who still drank directly from the Thames.[22]

By 1849, when cholera ravaged the city for the second time, anxiety about filth, pollution and disease seemed to outstrip the very limits of description. In this respect John Snow's cool analysis of the way in which the flushing of the sewers during that year had exacerbated rather than delimited the epidemic was untypical.[23] More expressive of what was now seen as an undeniable connection between a vilely polluted river and the threat of cholera were the lamentations of *The Times*. 'What was carried away in the sewers', the paper despairingly noted, 'was only transferred to the waters of a tidal river in the heart of the metropolis, there to be wasted in nothing but the production of disease.'[24] And the entire city, and not only the Thames, was now subject to this insensate pollution. 'Underground slaughterers', it was reported, 'red-armed and ankle-deep in putrid garbage, ply their loathsome trade. . . . Cartloads of half-putrid bowels . . . go steaming through the streets of the city, to be twisted into fiddle-strings on the cat-gut maker's wheel' and in 'pestiferous factories . . . the fat of animals' was 'converted into tallow . . . their blood into Prussian blue.'[25] By the early 1850s, during the respite between the second and third cholera epidemics, attacks on the state of the Thames and the water derived from it proliferated. Visitors to London were now said to avoid drinking company supplies out of a fear of dysentery and diarrhoea.[26] Sailors, who had formerly stocked up with Thames water before a long voyage, were frightened to do so.[27] Bad water was said to cause sickness, and sickness led irrevocably to pauperism and a decline in social discipline.[28] The leader of a parochial protest group contended that it 'was impossible to drink the water supplied to us through the pipes it tastes so badly',[29] while the chairman of another body claimed that working people endured water which would never be touched by the professional classes. 'We have found the public', he went on, 'very backward in coming forward.'[30] Was this apathy a form of self-protection? It may have been—a medical inspector in Bermondsey reported in 1854 that 'the clause of the Water Act, whereby tenants or landlords may be compelled to lay on water has never been put in force. There would be obvious injustice, it is felt, in forcing upon them such bad water.'[31]

Following the cholera crisis of 1854 the terms in which the pollution of the river was expressed remained intensely pessimistic, bordering at times on the cataclysmic. *The Times* continued to search for some means, any means, of cleansing the 'Cloaca Maxima'.[32] And *The Lancet* pictured the sea 'rejecting the loathsome tribute' and 'heaving it back again with every flow. Here in the heart of the doomed city, it accumulates and destroys.'[33] The continued spread of water-closets and the unrestricted flow of sewers into the inner-city river now received further official attention. 'Reform', warned a report on the recent outbreak of cholera, 'remains but imperfect and precarious while those river-side populations exercise a right of sewerage into the drinking water of London . . . the advantages to be gained from [modern house drainage] will suffer a serious counterpoise, if they can be purchased only at the cost of making the sewage-outfall into the river; if the change must be, from an unwholesome house to a polluted water-source; if that which would have been poison to inhale is to return as poison to drink.'[34] 'If the general use of water-closets is to continue and to increase', John Snow insisted, 'it will be desirable to have two supplies of water in large towns, one for the water-closets, and another, of soft spring or well water from a distance, to be used by meter, like the gas.'[35]

And yet, immediately before the crisis of 1858, there were those who claimed that both the river and the water taken from it were less repulsive than they had been a few years earlier. Francis Godrich, the medical officer for Kensington, recorded in 1857 that there had been a decline in dysenteric infection and that this was almost certainly due to the improved quality of company water.[36] As for Snow, he believed that 'the river in London was never in a better state . . .'. His only fear was that, if, as a result of major sewage works, the Thames might suddenly appear to become much clearer, the ill-informed would once again resort to drinking directly from it.[37] But events between July 1857 and September 1858 converted environmental deterioration into a vehicle for a series of debates on issues which were central to mid-Victorian political discourse. The parliamentary context was complex, with the Conservatives forming a brief minority government following a short-term crisis, triggered by the Orsini affair, within the Whig–Liberal alliance which dominated British political life between the fall of Peel and the Second Reform Bill.[38] The ideological and philosophical focus around which the debate on pollution was organised was 'local self-government'—the notion, deeply embedded in the consciousness of a majority of members that it was the locality, rather than the state, which must order and, crucially, pay for its own affairs.[39] But related to this were the questions of how the capital should be governed and the balance, on the one hand, between the City of London and the recently established Metropolitan Board of Works, and, on the other, between the Board and the national executive. Nor was this all, for, as the debate was to reveal, the Thames itself now came to be perceived in an explicitly 'imperial' context. The ramifications here were both practical and symbolic. Ministers, during an era of rampant colonial aggression, referred increasingly and rhetorically to the 'imperial' rather than 'national' Exchequer. And, as the crisis deepened, back-bench members also insistently depicted the Thames in these same extra-national terms. It was now widely acknowledged to be a national—and an imperial—humiliation that the very heart of the capital should be so vilely polluted, with the most ancient of parliaments subjected to environmental desecration.

The alarm was sounded by the Duke of Newcastle in July 1857, when he warned that 'the river was . . . like a vast sewer, and unless something was done before long to purify it, it would engender some frightful plague among the two and a half million who inhabited the metropolis.'[40] But full crisis was averted for nearly a year. Then, in the summer of 1858, there were suddenly 'strange stories flying of men struck down with the stench, and of all kinds of fatal diseases, upspringing on the river's banks'.[41] We shall see in a later chapter that these reports were indeed probably no more than 'stories' and that 1858 was a relatively healthy year, but during that traumatic summer, rumour inside and outside Parliament took a firm grip; and there were incessant demands for government intervention. But what could be done? There were those, like G W P Bentinck, member for West Norfolk, who castigated the capital and all its works. 'It was monstrous', he complained, 'that the richest city in all the world, which was surrounded by two of the richest counties in England should be always trying to relieve itself from the obligations that attached to itself alone.'[42] A S Ayrton, member for Tower Hamlets, took the opportunity to press for a total reform of metropolitan government. 'There was no escape from the dilemma', he insisted, 'except by giving the metropolis real and effective municipal institutions and so putting it on the footing in that respect of every considerable town in the kingdom.'[43] In a later debate Bentinck developed a slightly different variation on the theme of local self-government. 'The representatives of the metropolis', he said, 'had expressed their determination that the metropolis should not pay more than its fair proportion of the expense for purifying the Thames; so, on the other hand, the representatives of the country at large were determined that the country should not be taxed for metropolitan purposes.'[44] The issue was put in even more uncompromising terms by P Blackburn, representing Stirlingshire. 'The whole question was this', he said, '. . . the inhabitants of a very large town put an enormous quantity of dirt into their very fine river, and then they wanted the inhabitants of smaller and poorer towns to come and take it out for them.'[45] Not a penny should be spent to alleviate the capital's problems. All the more so since, Blackburn went on, 'the medical authorities admitted, though apparently with a feeling akin to regret, that the nuisance had not affected the public health to any appreciable degree'.[46] But it was Gladstone who made the most comprehensive case in favour of undiluted local self-government, 'discipline' and economy. London was not, and must not be allowed to be, an exceptional case. 'If it should be the opinion of that House', he urged, 'that London, the richest city in the world, should not bear the cost of draining its own river, as was done by less wealthy and less important places, what would be the future position of local government in London if that House interfered to do that which ordinarily ought to be done locally?'[47]

Despite the probable presence within the House of a clear majority in favour of local self-government in its purest form, it was the quasi-collectivist solution to the problem of pollution in London in 1858 which carried the day. That this was so owed little to the skills of ministers who stumbled from crisis to crisis, first denying then accepting partial responsibility for the state of the river and sometimes giving garbled accounts of existing administrative arrangements. Rather, it was the result of pressure from every part of the House that the 'national' and 'imperial' character of the river must now be acknowledged. When this set of ideas had taken hold, the Government found it easy enough to accept a compromise which could do little enough harm to the already blurred image and ideology of mid-nineteenth-century Conservatism. Vivid rhetoric depicting the potentially destructive results of inaction made a profound impact. 'The noblest of rivers', it was claimed, 'had been turned into a cesspool.'[48] It was 'a scandal before all Europe'.[49] 'God had given . . . a most magnificent river, and they had turned it into the vilest of sewers.'[50] 'They had built on the banks of the Thames a magnificent place for the Legislature but how could they direct the attention of any foreigner to it when he would be welcomed by a stench that was overpowering?'[51]

It was one thing to deploy such powerfully patriotic and grandiose language, but quite another to find a way through the political and institutional thickets which surrounded the administration of the Thames in the mid-nineteenth century. The Government did little more than point to the complexity of the problem. Disraeli, as Chancellor, admitted, without any great conviction, that the House had a 'moral responsibility . . . to do all in our power to prevent public disaster'.[52] Lord John Manners, as First Commissioner of Works, lamented that 'if the Government took upon themselves the main drainage of the metropolis at an enormous expense they could not in justice call upon the metropolis to pay for that drainage; while on the other hand, if the metropolis were called upon to do the work, it was but fair that they should do it in what manner it pleased. The responsibility must be imposed on the one party or the other.'[53] The executive was seeking 'to meet the evils that were supposed to flow from a double government and a divided responsibility'.[54]

A balance—between London and the Exchequer—must be rapidly arrived at. And here again the rhetoric of crisis played its part. 'If an invading army attacked the metropolis', Major-General T P Thompson, member for Bradford, told the House, 'it would not be urged that the provinces ought not to assist . . .'. For London was, after all, 'a collection of interests from the whole country'.[55] 'To say that this was not a national work, when men in that House, representing all parts of the country, loudly cried out against the evil . . . was manifestly pre-posterous.'[56] 'Parliament and the State contributed to the stench of the Thames', T S Duncombe, the member for Finsbury, contended; 'surely they ought to contribute towards its purification?'[57] 'The difficulty', it was insisted, 'arose from their having made a local which was essentially an Imperial question.'[58] Thus it was that the ideal of local self-government came to be eroded and modified. The Metropolitan Board of Works was entrusted with responsibility for the construction of a main sewage system; and the State, transcending localist dogma, was to provide financial guarantees.[59] What was unthinkable was that Londoners should be bled white for a scheme which was now so widely admitted to be unambiguously linked to notions of national and supra-national integrity.[60]

The crisis of the Thames in 1858, then, involved a partial re-evaluation of an important body of constitutionalist dogma—purist notions of local self-government, the role of the national executive in relation to governing agencies in the capital, non-intervention or no more than minimal intervention in municipal public works. It also stimulated reflection and debate on the relationship between the capital and the rest of the country. Historically, the image of London as a sink of iniquity, a gigantic parasite, sucking in the 'best quality' from the remainder of the country had been a powerful and a fertile one.[61] We must resist overstating the case but there can be little doubt that the crisis of 1858 modified that stereotype yet further. What deeply affected London, what degraded London, what threatened the 'nobility' of London—all this was now seen to have economic and cultural repercussions for the rest of the country.

But what, more specifically, of the roles of pollution and disease? Several members spoke during the debates of 'pestilence', 'plague' and 'calamity'. There was, in other words, continuity with conceptions of the nature and transmission of disease which had been current in the 1840s and earlier. But pollution and death were now also represented as dire punishments for those who might reveal themselves morally unwilling to take on the task of repurifying the capital. London was perceived as the potentially rotten heart of the body politic: and if London were to be fatally afflicted the rest of the country would almost certainly perish. All this had the effect, again, of unifying opinion in defence of distinctively metropolitan values. The Empire without: decay and rottenness within—these were the meanings and ideological rhetorics which were generated and deployed in the interest of social stability. To save the river was to consolidate the new urban–industrial order.

In 1858, then, there were interactions, at every level, between the social and the natural. We have seen vocabularies transposed, new imageries introduced into political debate and equations made between social and environmental stability. We have also noted that attempts were made to describe the threatening and the unknowable in terms sufficiently vivid and lurid to goad the Government into action. What may be detected from the early 1860s are new sets of environmental and political interactions as well as shifts in the rhetoric of pollution itself. The classic mid-nineteenth-century vocabulary of 'plague', 'pestilence', 'corruption' and 'decay' underwent gradual modification and, when cholera reappeared in 1866, it was depicted in terms which were more self-consciously analytic than fatalistic or apocalyptic. This may be explained partly in terms of autonomous scientific change but partly, also, in terms of wider social transformations. In a society less frequently disrupted by sudden fluctuations in the demand for exports, in which standards of living were rising slowly and the general (though not the infant) death rate declining, epidemiological and environmental crises were becoming less acute and less threatening to urban stability.[62]

An additional, dynamic aspect should also be noted. Environmental concern and anxiety never 'stand still'. Precisely what determines, within a given culture, what is environmentally acceptable, and how what is acceptable changes over time, are complex issues. Equally problematic are the rate at which and the mechanisms whereby the 'focus' of a pollution problem is transposed from one geographical area to another. Certainly, in this latter respect, many Londoners had come, by the mid-1860s, to believe that greater dangers were posed by pollution on the upper than the inner-city Thames. 'While London at this moment can boast of a water supply fully commensurate with the just demands of its mighty population', one observer wrote in 1859, 'it is, at the same time, threatened from the drainage, which its abundant supply of water so greatly facilitates, with a pestilential evil.'[63] What was being referred to here was uncontrolled sewage contamination of the Thames by other rapidly expanding urban areas. The problem, in other words, had shifted, but it had not been solved. 'Why the Londoners', it was asked in the early 1860s, 'do not rise up in arms against us provincials for remorselessly pouring our sewage down upon some of the water companies is a standing marvel.'[64] The 'provincials' had been urged to build their own sewage systems under the Health of Towns and Local Government Acts but this was legislation which had been framed with scant regard for the protection of the environment.[65] By the mid-1860s the position, as outlined by Robert Rawlinson, the prominent sanitary engineer and colleague of John Simon, was extremely serious. No community, he claimed, between Cricklade and Teddington Lock had prevented its domestic refuse from flowing into the river.[66] And with one or another form of the 'germ' or 'poison' theory of infection now gaining greater support it was widely expected that disease would 'flow' insidiously down the river and decimate the capital. 'The entire sewage of Windsor which is a drained and sewaged town, and is a sample of what the other towns will be shortly, goes out in one whole block, two yards wide and a foot deep ordinarily, bodily into the Thames.'[67]

John Simon was equally disconsolate. The completion of the main drainage system would certainly improve the quality of the water taken into the companies' reservoirs: but the manner in which the sewage was being disposed of above Teddington Lock was a danger to the health of every Londoner.[68] By the end of the decade the recurring dilemma—how best to combine the disposal of sewage with the protection of rivers—had still not been resolved. Attention was certainly now being more intensively concentrated on upper river communities but, as the Thames Conservators recorded in 1868, large volumes of sewage were still sometimes flowing illicitly into the inner river between Chelsea and the Isle of Dogs.[69]

But what of the overall quality of London's river? The Metropolitan Board of Works was, and would remain, sanguine. The London Bridge Sewer might still be in a disgusting state in the mid-1860s—with

steamboat passengers having to be protected by means of regular
chemical deodorisation of that part of the river—but, in general, the
Thames 'was not offensive'.[70] By 1867 fish, harbingers of at least a
degree of salubrity, were being spotted on the inner river, and this
improvement, the Board was at pains to point out, had not been
achieved at the expense of the health of soldiers said to have been struck
down by 'noxious vapours' at Woolwich near the massive outfall at
Crossness.[71] 'Further proof', it was reiterated in 1869, 'of the benefits
derived from the Main Drainage is to be found in the increased and
increasing purity of the river, of which the number of fish which
continue to be found, even in those parts of the river which were
formerly the most polluted, affords undoubted evidence.'[72] Even during
a hot summer, with lower than average rainfall, marine life continued to
revive: there would, it was said, 'be no return to the conditions of
1858'.[73]

There were nevertheless discrepancies between these accounts and
specialist evaluations of the water delivered to consumers in London.
There was a good deal of wariness, also, on the part of official investigators and metropolitan medical officers of health as to the possibility of
ever obtaining satisfactory supplies from a river which had been as
grossly polluted as the Thames. This was a theme which would recur
repeatedly between the mid-1860s and the end of the century, pre-
occupying governmental inquiries and deeply dividing the capital's
scientific élite. The resolution of the controversy—in what circumstances could a river ever be said to be irrefutably 'above suspicion'?—
both at the level of scientific speculation and lay intuition will be
considered in greater detail below; but to underestimate its significance
is to misunderstand the passion and animosity generated by river and
water pollution, as well as the 'water question' in general, during the
second half of the nineteenth century. That extreme scepticism was
justified was horrifically demonstrated by the sudden eruption of
cholera in the East End in 1866. The numerous investigations into that
epidemic directed attention to severe qualitative shortcomings and the
continuing unpredictability of patterns of supply in many sections of the
capital—the hopes expressed by John Snow, a decade earlier, that the
'floating population' on the Thames would never again drink raw water
from the Thames were now disappointed.[74] But pessimism appeared
to be refuted by the conclusions of a Royal Commission in 1868–9.
'Having carefully considered all the information we have been able to
collect', the commissioners stated, 'we see no evidence to lead us to
believe that the water now supplied by the companies is not generally
good and wholesome.' And yet an important rider, that if one or another
version of the germ theory were to be scientifically substantiated,
a less optimistic attitude would have to be adopted, reduced the

credibility of this position.[75] Nothing approaching bacteriological certainty had yet been achieved but an influential minority of those concerned with public health in London now championed a generalised version of the germ or 'poison' theory of disease, and were convinced that numerous serious outbreaks of disease had been directly transmitted via polluted river water.

By the early 1870s attention had been diverted towards another important socio-environmental problem—'domestic pollution' or the process whereby water delivered to consumers in a moderately safe state allegedly underwent serious deterioration in contaminated butts and poorly constructed receptacles. Given that domestic pollution was a fact of life in London throughout the nineteenth century, it was the ideological stance of medical officers and others which determined how it could be best explained and eradicated. One approach was to berate the poor for their ignorance and lack of hygiene, while another, diametrically opposed to it, was to argue that the evil would only be eliminated if constant supply were extended to every section of the community under municipal control. Here, as elsewhere, we see pollution, and the prevention of pollution, mediating fundamentally divergent attitudes towards the larger social and political order. 'A majority of the public', the medical officer for Paddington concluded in 1869, 'is not prepared for submitting to the conditions which a constant supply would necessitate.'[76] The alleged ineducability of the working class in hygienic matters was also noted by the medical officer for Kensington who claimed that older inhabitants developed a propensity for the distinctive taste of well-water and refused to drink more healthy company supplies.[77] Frank Bolton, who was appointed water inspector at the Board of Trade in 1871, adopted an unambiguous attitude towards the habits of the poorest consumers. 'Many of the cisterns, tanks and butts', he reported in 1875, 'for containing water in small tenement houses in the Metropolis are in a disgusting and filthy state. An opportunity for inspecting these presents itself when travelling on some of the Metropolitan and suburban lines. Cisterns may be seen without lids, full of rank and decaying vegetation which on closer examination would show more or less organic deposit, and under the microscope would be found to abound in infusorial life.'[78] Bolton was also convinced that the 'water companies are frequently blamed for delivering impotable water, when if the true delinquent were sought it would be found to be the water consumer himself whose lack of attention to his cisterns and filters has created the evil of which he complains.'[79]

Those who favoured more and cheaper water under municipal control had an interest in providing equally lurid evidence. 'The means of storage'—this was in Kensington in 1870—'is both bad in its character and insufficient in amount. Where, therefore, an abundant supply is most needed, it is too often lacking; and in many cases a common cistern supplies all the wants of the household, even including the service of the water-closet.'[80] '[The] liquid', it was observed in 1872, 'is frequently a solution of dropped leaves, bits of plaster, old boots and various kinds of rubbish.'[81] Whatever solutions reformers might advocate to improve the position of working-class consumers in London in the 1870s, there could be no denying the existence of an indissoluble link between the quantity and quality of delivered water. Pollution of the Thames and of supplies derived from it had by this time been significantly reduced, but the absence of large amounts of cheap water on constant supply invariably led to false economies and hence to risks of water-transmitted or water-related infection. The companies and their supporters assumed that possession of an effective and hygienic domestic storage system was part and parcel of being a responsible customer. The fact that working-class consumers moved much more frequently than the middle and upper classes, that the rooms in which they lived were ill-equipped to store adequate amounts of water for immediate use, and that the areas in which *as a class* their housing was located were insalubrious and unlikely to provide access to clean butts and tanks—all this was ignored. Poverty, in other words, could lead to pollution and ill-health at precisely the same time as administrative action was reducing the generalised filth of the river. Here again we see the extent to which the incidence and impact of environmental deterioration, in a river or in the individual home, was socially and politically determined. Access to cleanliness was at root a function of wealth and income and the denigration of working people said to be too lazy or too ignorant to ensure that the water which they drank was correctly stored constituted little more than blatant legitimation of the *status quo*.

If the issues which have been discussed in the preceding paragraphs may be loosely defined as 'practical', though still, at the deepest level, socially and politically determined, other concerns in the mid-1870s were primarily theoretical and scientific. The most important of these has already been mentioned, namely the debate about whether the Thames, which had been subjected to such astonishingly high levels of pollution between roughly 1820 and 1860, would ever again constitute a safe source of supply for the metropolitan water companies. This was a question to which the Rivers Pollution Commission turned its attention in 1874, arriving at conclusions which were anything but reassuring. Thames water might now finally be 'free from offensive taste or odour' but there was little else to be said in its favour. Impregnated by the sewage of upper-river communities and by chemical washings from agricultural land, it was further harmed by the impact of public bathing, sheep washing and 'putrid carcasses'. Flooding and sudden fluctuations in temperature increased the propensity to 'organic' pollution and filtration did little to reduce the effects of sewage on potability. Only one source within the capital, the Commission conceded, was beyond reproach and this was the deep well water delivered by the Kent Company. 'The supply of such water', it was concluded, 'either softened or unsoftened, to the metropolis generally would be a priceless boon, and would at once confer upon it absolute immunity from epidemics of cholera. We are very decidedly of opinion that the metropolitan companies should receive, from Your Majesty's Government, sanction for increase of capital, only on condition that such capital shall be extended on works necessary for the supply of this palatable and perfectly wholesome beverage.'[82]

The reports to the Registrar-General on company-delivered water during the 1870s and early 1880s by Edward Frankland, the chemist, highly influential specialist on public water supplies and driving force behind the Rivers Pollution Commission, confirmed these conclusions, or prejudices as they were claimed to be by those who disagreed with the belief that any supplies which had been subjected to 'previous sewage contamination' were unfit for human consumption.[83] During December 1876 Frankland reported that the Thames had been laden with 'organic matters of the most objectionable origin' but that supplies taken from it had still been distributed in very large quantities by the West Middlesex Company. In that same year only three out of the seven concerns were said to have delivered their water in an 'efficiently filtered condition'. In 1877 each of the companies was deemed to have distributed supplies which failed to satisfy the criteria laid down by the Rivers Pollution Commission and in 1878 floods had swept down large quantities of sewage which had dangerously harmed metropolitan drinking water. During 1880 the water distributed by the companies had only been 'fit to drink' in three months out of twelve; and in 1881 most of it had been unsafe in January, February, March and December.[84]

To what extent were Frankland's theories influenced by, or themselves an elevated reflection of, opinion among medical officers and others concerned with health and environment in London at this time? The evidence suggests that Frankland's purist position did not command widespread support—indeed, only the medical officers of Lambeth and Wandsworth in the 1870s and 1880s appear to have had expectations of the companies, and of the Thames, that were as demanding as Frankland's own. In 1880 the medical officer for Lambeth warned that, during the autumn in particular, large volumes of sewage were poured into the Thames by upper-river communities, thus making it highly likely that diarrhoeal infection would periodically afflict

Londoners.[85] The view from Wandsworth was even bleaker. 'The water supply of London', the medical officer insisted, 'is a disgrace to the sanitary legislation of the present day.'[86] John MacDonagh, medical officer for the Clapham district, took up an unambiguously Frank-landite stance. 'The water companies one and all', he wrote in 1881, 'spare no expense or trouble in their endeavours to purify Thames water. But they cannot do what is impossible; they cannot make it fit for human beings to drink.'[87]

But the epidemiological record seemed to cast doubt on the validity of Frankland's exceptionally strict standards for a safe source for large-scale supplies of drinking water. 'If I am drinking Thames water', Frank Bolton commented, 'I should be satisfied with it; and I do not think I should change my residence to get into the Kent district for the sake of the water. Some people like one, and some like the other.' He supported this position by arguing that the statistics indicated that complaints by the public about quantity greatly outnumbered those to do with quality.[88] (How valid such figures actually were, however, must be open to doubt since, as has already been seen, inadequate domestic supplies themselves indirectly led to precisely the same social and epidemiological problems as water that was already polluted on delivery.)

And yet the tone of those who championed the Thames as a source for the capital during the later nineteenth century, and repeatedly attacked Frankland's preoccupation with 'previous sewage contamination', verged at times, on the euphoric. 'It is wonderful', the medical officer for Marylebone enthused in 1883, 'how constant the composition of the public water supply is maintained; the variations in the composition being entirely dependent on season; the impurities showing an increase in rainy seasons, a decrease in dry . . . the wholesomeness of the water, as delivered from the mains, is a fact too well established to be profitably discussed.'[89] It was, though, precisely these 'variations', predictable or otherwise, which most alarmed Frankland and his supporters. A more convincing though still overstated anti-Franklandite critique was marshalled by Louis Parkes, the highly active medical officer for Chelsea. 'It was', he wrote in 1887, 'a very remarkable fact that although the specific poisons of enteric and other fevers are constantly passing into the Thames but a few miles above the water companies' intakes yet no evidence has ever been forthcoming of disease traced to the quality of the water supplied by any of the Thames companies. . . .'[90] Following a number of unprecedentedly optimistic reports in the mid-1880s, by the early 1890s Frankland was again expressing himself in uncompro-misingly gloomy terms. 'The storage, subsidence and filtration plant of these companies who purvey river water were quite inadequate to deal with the exceptional pollution: and water of such bad quality, as regards organic matter in solution, has rarely been delivered by these Companies during the last 26 years. Even the Chelsea and East London Companies, the storage capacity of whose reservoires is from twice to five times that of the other river water companies, cannot at all times exclude foul flood water.'[91] This time Louis Parkes agreed, and in an article entitled 'The Air and Water of London: Are They Deteriorating?' claimed that on several occasions the water had not been 'fit for dietetic use, that is to say *not safe to drink*'.[92] Lambeth also kept up its protests and made (unsuccessful) representations to the Board of Trade and the London County Council.[93]

But the Royal Commission of 1893–4 remained unmoved. 'We are strongly of opinion', it stated, 'that the water, as supplied to the consumer in London, is of a very high standard of excellence and of purity and that it is suitable in quality for all household purposes . . . we do not believe that any danger exists of the spread of disease by the use of this water, provided that there is adequate storage and that the water is efficiently filtered before delivery to the consumers.'[94] The Commission felt able to go even further. 'Every medical witness that has appeared before us, whether his general feeling was favourable or unfavourable to the water, has told us unhesitatingly that he knows of no single instance in which consumption of this water has caused disease.'[95] Such conclusions certainly had the support of those like Sir William Crookes, a Fellow of the Royal Society, and water analyst for the companies, who believed that, in statistical terms, it was probably then more dangerous to breathe London air than to drink London water.[96] Most important of all, Edward Frankland himself, impressed by recent developments in bacteriology and filtration technology, had begun to modify his apparently unshakeable pessimism.[97] It was, though, the bacteriologist, Sims Woodhead, who issued a warning against excessive optimism. 'The water supply of the largest city', he said, 'should be as much above suspicion as Caesar's wife, but so long as it contains a large quantity of animal organic matter, and so long as pathogenic or even saprophytic organisms, other than those usually found in water, can make their way into it, it cannot be looked upon as a desirable water supply. . . .'[98] This coincided with the view of Shirley Forster Murphy, the first medical officer to the London County Council, who weighed up the hypothetical dangers of drinking water derived from the Thames and Lea with those associated with crossing the street. Using an argument analogous to that deployed by Crookes he arrived at quite different conclusions, and went on to argue that anything approaching absolute safety would only be achieved when the water supply had been fully municipalised. As so often in the past, complacency was now punctured by serious outbreaks of water-transmitted disease. In 1894 and 1895 there were epidemics of a diarrhoea-like infection in the Strand district which appeared to be jointly attributable to contaminated oysters and inefficient filtration.[99] The process of filtration itself now came under close scrutiny and the possibility of 'microbes' 'breaking through' into the water mains was widely canvassed.[100] For some, though not as we shall see for everyone, the application of bacteriological knowledge to water treatment now played an important role in re-establishing confidence in the Thames and the public supplies derived from it. Thus in 1899 George Newman, medical officer for Clerkenwell, was convinced that since water delivered by the New River Company contained no more than eight to ten microbes per cubic centimetre, as compared with a figure, for raw Thames water, of between 1500 to 2000 microbes, 'infection through the water supply may be considered as extremely unlikely, if not impossible'. 'The drinking water of the parish', he continued, 'may be looked upon as of very pure quality as regards bacteria. Probably not one of the 92 recorded cases of Typhoid Fever was derived from our water supply. There is of course the further argument that if the water supply was the cause of our high typhoid rate, we should expect the disease to be more widespread. For the water supply is common to the entire population.'[101] Where Newman worked through the evidence to arrive at credible, though over-optimistic, conclusions, others grossly exaggerated the safety of Thames water. 'For upwards of thirty years', wrote Arthur Shadwell, 'no sickness whatever has been traced to the London water, and the steady diminution of water-borne disease is a fact which no academic discussion can over-ride.'[102] William Crookes went even further in his defence of the companies and offered to drink water directly from the river. 'I do not think', he said, 'that one would take harm.'[103] Fortunately for him, he resisted this inadvisable urge towards self-experimentation.

Surveying the state of the river in the later 1880s the medical officer for Lambeth described what he saw and felt in the following terms. 'Lambeth is a riparian parish, the waters of the Thames wash its shores, and the waters receding leave that portion of the Parish, which at high tide is mythical, at ebb tide a sad reality. Can anyone who has an interest in the health of those who people the riverside districts view the masses of black slime, pregnant with disease, that disfigure the bank, without the consciousness that such an influence must be produced which is baneful to the public health in the present, and may have far-reaching effects in the future?'[104] He went on to hold out the hope of 'a river no longer redolent with chemicals and viscid with the deodorised effluent from the presses of Crossness, but a stream whose tide, crisp from the sea, shall bear in the flow, a store of ozone to replenish the impoverished air, and again in the ebb, draw back in its waters, the oxidised impurities of the atmosphere.'[105]

If there had been a shift between the middle and the end of the nineteenth century from cataclysm to partial control, from a belief that the unprecedentedly polluted river might actually destroy the city, to a degree of confidence in the new insights of bacteriology and water 'science', it was not reflected in observations such as these. For this vision was a holistic one, according to which the river was perceived, as it had been two generations earlier, as symbolic of health and purity. To allow it to deteriorate and to do nothing to conserve its aesthetic qualities was both dangerous and irresponsible. What may be detected here is a commitment to a quasi-eighteenth-century ideal of the river— any river—in its pristine state, as a harbinger of generalised health. The bracing air flows along the waterway, ozone from the sea and light breezes from the countryside revitalise the foul and torpid city. The implications are moral as well as environmental. But, by this juncture, such views were untypical. Our analysis of pollution, and the language of pollution, has revealed that near helplessness had been replaced by a commitment to intervention. The unbridled, at times almost nihilistic, lamentations of *The Times* and the medical press of the 1840s and 1850s had given way to the measured tones of the annual reports of the metropolitan medical officers of health and numerous official investigations into the water supply of the capital.

There remain, though, two important questions: the extent to which transformations in language reflected a real, measurable reduction in levels of pollution; and the ways in which the movement from a sense of impending calamity to a conviction that river pollution was manageable was mirrored in or determined by external sociopolitical change. If quantitative data on the state of rivers in the late twentieth century are open to a wide range of interpretation, the material which has survived from the nineteenth is even more ambiguous and potentially misleading. Having said that, one must immediately add that the epidemiological record, and, more specifically, mortality and morbidity from typhoid, indicates that the Thames, and the water derived from it, was much less unhealthy in 1900 than it had been in 1850. Here, again, though, there is need for qualification, for expectations of what constituted acceptable levels of health had themselves been raised and refined. Whether what was by then deemed satisfactory in terms of public health and environmental quality was expressed in scientifically analytical terms or more effusive and overtly morally loaded language was less significant than the fact that new norms had in fact been established. (One difference between the 'optimists' and the 'pessimists' was that, while the former tended to be satisfied with the existing consensus, the latter were always ready to initiate a new round of 'upward' negotiation.)

If the theoretical and methodological issues surrounding the evaluation of pollution in the past are complex, those to do with the connections between pollution and the social order lie at the heart of a genuinely social history of science and environment. On a number of occasions in this chapter there have been indications that there were congruencies and mediations between the natural and the social. The moral panic over river pollution and the crisis of the inner city in the 1840s and the 1850s; the appropriation of the river as a symbol of nationalism and imperialism in 1858; the simultaneous trends towards a degree of social cohesiveness and river management as standards of living slowly rose and rampant in-migration became less disruptive in the early 1870s—in each of these instances we may discern compelling structural linkages between river pollution and the sociopolitical domain.

1   *Royal Commission on the Water Supply of the Metropolis* PP 1828:IX:61.
2   *Ibid*, 61–2 and 122–3. See, also, the revealing table on catches of salmon between 1794 and 1821 in Leslie B Wood, *The Restoration of the Tidal Thames* (Bristol, 1982), 18.
3   *The Times*, 20 August 1827 cited in *RC Water Supply Metropolis*, 200.
4   *RC Water Supply Metropolis*, 199.
5   *Ibid*, 62.
6   *Ibid*, 168.
7   *Select Committee on the Health of Towns* PP 1840:XI:Q 3452. The growth of water-closet provision by the end of the first third of the nineteenth century is also documented in 'Reports on Plans for the Main Drainage of the Metropolis, as Submitted to the First Commissioner by the Metropolitan Board of Works' PP 1857(2):XXXVI:6(n) and C E Saunders, 'Legislation for the Purification of Rivers and its Failures', *Trans. Soc. Med. Officers of Health* (1886–7) 76.
8   *RC Water Supply Metropolis*, 114, evidence of John Mills.
9   *Ibid*, 149, evidence of Dr Pearson.
10  Reduced concern over the water question during the 1830s is noted by Anne Hardy, 'Water and the Search for Public Health in London in the Eighteenth and Nineteenth Centuries', *Med. Hist.* **28** (1984) 263 and by D Lipschutz, 'The Water Question in London' *Bull. Hist. Med.* **42** (1968) 510–25. I am in general agreement but my explanation is a different one.
11  Joseph Fletcher, 'Historical and Statistical Account of the Present System of Supplying the Metropolis with Water', *J. Stat. Soc.* **vii** (1845) 159 and 161.
12  *Select Committee on Metropolitan Sewage Manure* PP 1846:X:Q 1091.
13  Francis Sheppard *London 1808–1870; The Infernal Wen* (1971) 247–53 and

377–83. M J Cullen, *The Statistical Movement in Early Victorian Britain* (Brighton, 1975) is crucial for an understanding of the crystallisation of specific areas of reforming activity in this period.

14  On John Martin and his 'projects' see Mary L Pendered, *John Martin, Painter* (1923) 196, 213 and 303–4; Thomas Balston, *John Martin, 1789–1854: His Life and Works* (1947) 122 and 125–8; and Francis Klingender, *Art and the Industrial Revolution* (1947) 104–9.

15  *Select Committee on the Sewers of the Metropolis* PP 1834:XV:371–5.

16  *SC Metropolitan Sewage Manure*, 1846, Q 1064.

17  *Select Committee on the Thames Embankment* PP 1840:XII:Q 239, evidence of John Walker.

18  *Annual Report of the Metropolitan Board of Works* PP 1889:LXVI:15–18.

19  *Royal Commission on the Improvement of the Metropolis* PP 1844:XV:154.

20  *Select Committee (H L) Supply of Water to the Metropolis* PP 1840:XII:Q 309, evidence of George Glasier.

21  *Royal Commission on the State of Large Towns and Populous Districts* PP 1844:XVII:Q 85, evidence of T Clarke, Professor of Chemistry, University of Aberdeen.

22  *Select Committee on the Health of Towns* PP 1840:XI:Q 729, evidence of M F Wagstaffe, surgeon.

23  John Snow, *On Cholera*, edited by Wade Hampton Frost (New York, 1936) 136–7.

24  *The Times*, 21 July 1849.

25  *Ibid*, 2 October 1849.

26  *Report of the General Board of Health on the Supply of Water to the Metropolis* PP 1850:XXII:Qs 724–5, evidence of Hector Gavin.

27  *Ibid*, Qs 730–1, evidence of Robert Bowie.

28  *Select Committee on the Metropolis Water Bill* PP 1851:XV:Qs 4844–54, evidence of John Challice.

29  *Ibid*, Q 5063, evidence of E Collinson.

30  *Ibid*, Q 6512, evidence of William Yates Freebody.

31  *Report of the Medical Council in Relation to the Cholera Epidemic of 1854* PP 1854–5:XLV:112.

32  *The Times*, 14 September 1854.

33  Cited in *MBW Report* (1889) 7.

34  *Report on the Last Two Cholera Epidemics of London as Affected by the Consumption of Impure Water* PP 1856:LII:368–9.

35  John Snow, *Medical Times and Gazette*, 20 February 1858, 190.

36  *Report of the MOH: Kensington* (1857) 31.

37  John Snow, *Medical Times and Gazette*, 20 February 1858, 191.

38  The most stimulating study of this complex period is J R Vincent, *The Formation of the Liberal Party 1857–1868* (1966): but see also Norman Gash, *Aristocracy and People: Britain 1815–1865* (1979), 250–83. For the Orsini affair see Jasper Ridley, *Lord Palmerston* (1970), 479.

39  Local self-government is an insistent leitmotif in Lambert, *Sir John Simon 1816–1904 and English Social Administration* (1963) but the account in Geoffrey Best, *Mid-Victorian Britain 1850–1875* (1971), 35–45 is also excellent.

40   *Hansard,* third series, **cxlvii** col 5.
41   R Barnes, 'Is the Thames Pernicious?', *J. Pub. Health and San. Rev.* **iv** (1858) 142.
42   *Hansard* **cli** col 437.
43   *Ibid,* col 438.
44   *Ibid,* col 578.
45   *Ibid,* col 577.
46   *Ibid,* cols 1165–6.
47   *Ibid,* cols 876–7.
48   *Hansard* **cl** col 2113. Capt. C E Mangles, member for Newport, Isle of Wight.
49   *Hansard* **cli** col 31. S Warren, member for Midhurst.
50   *Ibid,* col 29. Capt. C E Mangles.
51   As reported in *MBW Report* (1889) 12. Capt. C E Mangles.
52   *Hansard* **cli** col 441.
53   *Ibid,* col 37.
54   *Ibid,* col 36.
55   *Ibid,* cols 1166–7.
56   *Ibid,* cols 576–7. W Cox, member for Finsbury.
57   *Ibid,* col 1168.
58   *Ibid,* col 573. W Roupell, member for Lambeth.
59   David Owen, *The Government of Victorian London 1855–1889: The Metropolitan Board of Works, the Vestries and the City Corporation* (edited by Roy MacLeod, Cambridge, MA, 1982) chapters 2 and 3.
60   *Hansard* **cli** col 577. J Locke, member for Southwark.
61   The demographic and social parameters are provided in E A Wrigley, 'A Simple Model of London's Importance in Changing English Society and Economy', *Past and Present* **37** (1967) 44–70. But see also Lawrence Stone, *The Crisis of the Aristocracy* (Oxford, 1965) 386–98 and, for urban–rural relationships in general, Raymond Williams, *The Country and the City* (1973). P J Corfield, *The Impact of English Towns 1700–1800* (Oxford, 1982), chapter 5, is also highly relevant.
62   There has, however, been a partial re-evaluation in recent years of 'mid-Victorian prosperity'. R A Church, *The Great Victorian Boom* (1975) surveys the literature and points to great regional diversity, while, for London, Stedman Jones, *Outcast London: A Study in the Relationship between Classes in Victorian Society* (Oxford, 1971) chapter 13 depicts very serious economic difficulties in the 1860s. See, on a similar theme, Bill Luckin's 'Evaluating the Sanitary Revolution: Typhus and Typhoid in London 1851–1900' in R Woods and J Woodward (eds), *Urban Disease and Mortality in Nineteenth Century England* (1984) 102–19. In terms of social *stability,* however, the 1850s and 1860s were greatly less traumatic than the 1830s and 1840s.
63   John Strang, 'On Water Supply to Great Towns: Its Extent, Cost, Uses and Abuses', *J. Stat. Soc.* **xxii** (1859) 234.
64   J N Radcliffe, 'The State of Epidemic, Epizootic and Epiphytic Disease in Great Britain, 1861–2', *Trans. Epidemiol. Soc.* **i** 401.

65   *MBW Report* PP 1865:XLVII:10.
66   *Select Committee Thames Navigation Bill* PP 1866:XII:Qs 2774–5.
67   *Ibid*, Q 2583.
68   *Royal Commission on Water Supply* PP 1868–9:XXXIII:Q 2814.
69   *General Report of the Conservators of the Thames* PP 1868–9:L:638.
70   *MBW Report* PP 1865:XLVII:9.
71   *MBW Report* PP 1867:LVIII:10–11.
72   *MBW Report* PP 1868:LI:11.
73   *Ibid*.
74   *The Lancet*, 4 August 1866, 130.
75   *RC Water Supply* (1868) cii.
76   *Report of the MOH: Paddington* (1869–70) 16.
77   *Report of the MOH: Kensington* (1875) 30–1.
78   'Metropolitan Water Supply: Report by Lt-Colonel Frank Bolton, Water Examiner under the Metropolis Water Act, 1871' PP 1875:XXXI:266.
79   F Bolton and P A Scratchley, *The London Water Supply* (second edition, 1888), 13.
80   *Report of the MOH: Kensington* (1870) 45–6.
81   *Report of the MOH: St Giles* (1872) 4–5.
82   *Rivers Pollution Commission: Sixth Report: Domestic Water Supply of Great Britain* PP 1874:XXXIII:624–5.
83   This issue is well dealt with by Christopher Hamlin, 'Edward Frankland's Early Career as London's Official Water Analyst, 1865–1876: The Context of "Previous Sewage Contamination"', *Bull. Hist. Med.* **56** (1982) 56–76.
84   This section is based on Frankland's reports in PP 1877:XXXVIII: 122–5; 1878:XXXVII(Pt 1):261; PP 1878–9:XXVIII:178; PP 1881: XLVI:431; and PP 1882:XXX(Pt 1):209.
85   *Report of the MOH: Lambeth* (1880) 33.
86   *Report of the MOH: Wandsworth* (1880) 12.
87   *Report of the MOH: Wandsworth (Clapham district)* (1881) 65.
88   *Select Committee on London Water Supply* PP 1880:X:Qs 2060 and 2089–91.
89   *Report of the MOH: Marylebone* (1883) 135.
90   Louis Parkes, 'On Water Analysis', *Trans. San. Inst. GB* **ix** (1887–8) 385.
91   *Fifty Third Annual Report of the Registrar-General* xliv.
92   Louis Parkes, 'The Air and Water of London: Are They Deteriorating?' *Trans. San. Inst. GB* **xiii** (1892) 66.
93   *Report of the MOH: Lambeth* (1892) 37–8.
94   *Royal Commission on the Water Supply of the Metropolis* PP 1893–4:XL(1):73.
95   *Ibid*, 69.
96   *Ibid*, Q 10 477.
97   *Ibid*, Q 4471.
98   *Ibid*, Appendix C70 505.
99   *Report of the MOH: Strand* (1896) 139.
100  *Report of the MOH: Stoke Newington* (1896) 70–1 and *Report of the MOH: Clerkenwell* (1896–7) 115.
101  *Report of the MOH: Clerkenwell* (1899) 26–7.

102   Arthur Shadwell, The London Water Supply (1899) 52.
103   *Royal Commission on Water Supply within the Limits of the Metropolitan Water Companies* PP 1900:XXXVII(Pt 1):Q 21 659.
104   *Report of the MOH: Lambeth* (1886) 17.
105   *Ibid*, 19.

# 2  Treating the Water

The process of filtration, or what some nineteenth-century observers characterised as a 'barrier' which might prevent the delivery of foul and dangerous river water, is the main theme of this chapter. Attention is given to technical change, intercompany differentials and official investigations which sought to goad the companies into providing more wholesome water. The unreliability of the available evidence makes it exceptionally difficult to arrive at convincing generalisations about the efficiency of any given concern in terms of the interactions between the subsidence of raw river water in reservoirs, the depth and make-up of the filtering medium, and the rate of filtration. Indeed, a theme which will recur is that official 'rankings' during the first 30 years of our period were marred by disagreements over how the efficacy of filtration, as well as the total system of which it was a part, should be evaluated. (There is an intriguing parallel here with the divergent reasons put forward by individuals and groups during the same years to explain why a given phenomenon should be deemed an environmental threat.)

After 1870, the year of an important report by Netten Radcliffe for the Medical Office of the Privy Council, the quantitative material becomes more reliable. Paradoxically, however, debate now began to centre more on the underlying principles of filtration—why, precisely, the process was as it was and did what it did—than on such matters as the acreages devoted by each of the companies to reservoirs and filter beds. But there was now agreement that, among the concerns deriving the great bulk of their supplies from the Thames, the West Middlesex and Grand Junction Companies were the most efficient. From the early 1880s onwards attention was directed towards the implications for water purification of new insights derived from bacteriology. It might be assumed that the assurance that filtration had worked moderately well in the immediate past was explicable in micro-organic terms would have strengthened confidence in the treatment of water in London. But this was not always the case. Indeed, there was, as we shall see, a sense in which both the language and the imagery generated by the new science raised novel and, for some, unwelcome uncertainties. Thus it was that residual pessimism in relation to filtration modified and complicated the general movement from 'cataclysm' to 'partial control' which has been described in Chapter 1.

It is tempting to seek out symmetries in relation to filtration at this time and to argue that bacteriologists were more confident about the possibilities of finally eliminating water-transmitted disease, than those trained in chemistry. But such patterns are not in fact fully supported by the evidence. All that can be said with any degree of certainty is that by the very end of the nineteenth century the science of chemistry still carried within itself imageries and ideologies based on an intermittent preoccupation with sewage pollution and its underlying sociopolitical correlate: disorder within an urban–industrial environment.

There was little agreement over the role or effectiveness of filtration in London in the mid-nineteenth century. James Simpson, who considered himself the doyen of water engineers within the capital, had devised the first operational filter bed for the Chelsea Company between 1825 and 1834.[1] He had later been employed by the Lambeth Company and completed works for them by 1841.[2] Simpson preferred a very deep filtering medium—eight feet of gravel, shells and sand for the Chelsea Company, seven feet of gravel and sand for the Lambeth[3]—and he had no doubt, he told a select committee in 1851, 'that there had ever been a more perfect system of filtration ever introduced in this country'.[4] Simpson might be an enthusiast for water purification but others were less committed. Thomas Wicksteed, engineer for the East London Company, admitted in the same year that, although he was in favour of filtration, his directors were not. They believed that subsidence in well-constructed reservoirs was sufficient and that filter beds were an uneconomical luxury.[5] As for W C Mylne, engineer at the New River Company, he had been under pressure for some time both from his directors and from the general public to purify water tainted by the sewage of Hertford and other towns. But he remained unconvinced. First, there was the question of cost—it was essential, he contended, that at least one eighth of the entire filter surface should be undergoing cleansing at any one time and, for the New River Company, this would have involved an outlay of nearly £70 000 at annual interest payments of nearly £3500. Mylne retained an extraordinary confidence in the company water and claimed to drink a glass of it first thing every morning in his dressing-room.[6]

Other evidence to that same committee, which prepared the way for the Metropolis Water Act of 1852, revealed a wide diversity of attitudes towards the need to subject company supplies to filtration. One academic witness praised the quality of the river water above Hammersmith, argued against growing contemporary commitment to covered reservoirs but insisted on controlled filtration. But he was still willing to drink water impregnated with 'animaliculae' which were visible under a microscope.[7] Here the definition of what constituted a 'safe' supply was clearly determined by an exceptionally individualistic conception of the relationship between water and health. John Simon's views on filtration were unambiguous. Every company which did not filter its water must be required to do so immediately. The difficulty, though, both for Simon and others, was whether the water concerns should be compelled to regulate themselves or be subjected to external control.[8] And here we may detect the beginnings of a conflict—between capitalist individualism and municipal or state-backed collectivism—which would have a profound effect not only on filtration but on all other aspects of the 'water question' in London during the second half of the nineteenth century.

'The effect of ordinary filtration through sand', it was said in 1851, 'is very decided on Thames water. The river water can thus be easily obtained, unless in certain exceptional circumstances, entirely free from suspended matter, or chemical impurities.'[9] But precisely how unreliable such judgments were would be demonstrated by less optimistic assessments, and by contradictory accounts of the impact of filtration and subsidence on the supplies of individual companies drawing their water from the Thames. Some of these 'rankings' of the private water concerns in the period under discussion certainly deserve serious consideration. But others were less dependable and suggested that those who compiled them were in fact making as strong a political case as possible against the companies and in favour of municipalisation. That they felt constrained to deploy quasi-scientific data in this way was itself a condemnation of the companies' continuing refusal to put their own house in order or to allow themselves to be monitored by an external body.

The legislation of 1852 demanded that the companies store water taken from the Thames and the Lea in covered reservoirs if these were located within five miles of St Paul's and to subject it to filtration before delivery to the public. But technical and legislative enforcement were little more than rudimentary and official investigations continued to provide testimony both of gross insalubrity and of differentials between the companies. Thus, according to the Committee of Scientific Inquiries in 1854, 'the Chelsea water shows a much greater amount of dissolved impurities but (apparently as a result of filtration) far fewer visible forms; while in the Southwark and Vauxhall water this evidence of unfiltered contamination reaches its highest degree, revealing to the microscope, not only swarms of infusorial life, but particles of undigested food referable to the discharge of human bowels'.[10]

In 1856 the General Board of Health attempted, though with no more than minimal success, to discover what was best and worst in metropolitan waterworks technology. The major weakness in terms of more effective filtration was a lack of uniformity, and this was rooted in a system marked by 'no . . . plan of action for guidance' and superintended by engineers holding 'very different views'.[11] There were profound weaknesses within each of the companies as well as variations between them but, even worse, the criteria by which such judgments were made remained inexplicit and intuitive rather than rooted in close observation of day-to-day practice or controlled experiment.

Nevertheless the Board did feel able to point to the most and least safe company supplies and to technologies which may have explained genuinely striking differentials. Thus, of the five concerns drawing all their supplies from the Thames, the Chelsea was considered to be the most efficient and to possess works of a 'substantial and excellent kind'. These consisted of subsidence reservoirs and over three acres of filter beds. The filters themselves were eight feet deep and the rate of filtration—already recognised as an important index—approximately two gallons per square foot per hour.[12] The water supplied by the Grand Junction Company was less wholesome, but here also there had been progress—the filtration area had been tripled since 1850: the depth of sand and gravel had been increased from four to eight and a half feet and the filtration rate was three gallons per square foot per hour.[13] At the other extreme, the West Middlesex Company, which had used no filtration at all in 1850, possessed, by 1856, over one and a half acres of beds. But the rate was as high as six gallons per square foot per hour and the Board judged its water to be 'by no means fit for domestic use'.[14] This was harsh criticism but the Lambeth Company fared little better. Demand for its water had increased rapidly but the total filtration area—no more than 32 000 superficial feet—was wholly inadequate.[15]

The Board of Health report, then, did little more than seek to establish whether there had been gross breaches of the legislation of 1852. There were hints that interactions between every component of waterworks technology and practice—the source for raw river water, the state of the Thames when identifiable supplies had been taken into the reservoirs, the period allowed for subsidence and the method and rate of filtration—were thought to be as important as any single, discrete process. But crucial issues in what would later become the 'science of filtration' (the depth of the filter and the materials from which it was constructed) were left largely unexamined. Thus there was no questioning of why the Southwark Company made use of beds five and a half feet deep, the Lambeth seven, and the Chelsea eight; nor of materials which ranged from gravel, shells and sand, to gravel and sand, to Harwich sand. No explicit linkages were made between the depth of the filter, the material from which it was made, and the rate of filtration; or between the filtering medium and the colour of water on delivery. Lack of contact between the water concerns was deplored; but repeated statements by company engineers that 'tradition' played a crucial role, namely that water had always been filtered in a given way and that there had been very few complaints, were not queried. Water filtration was assessed more in terms of largely unenforceable legal requirements than a body of applied theory or in relation to the epidemiological record.

It might have been expected that, following the cholera epidemic which struck so savagely at the poorest sections of the East End in 1866, both the theory and the practice of water purification would have undergone substantial change. But this was not the case. The companies repeatedly denied links between poorly treated water and the spread of disease and only those public health officials, like William Farr and Edward Frankland, who had been closely involved with the outbreak insisted that the supplies of the East London Company had been almost wholly responsible for the appallingly high death rate. The dominant investigative attitude towards what was still considered to be the companies' private domain continued for the rest of the decade to be tentative, empirical and legalistic. It was only in the 1870s, when the full implications of the 'poison' or embryonic germ theory of disease had been more widely publicised and the connections between substandard supplies of water and the transmission of infections like cholera and typhoid more fully understood, that epidemiological insight revealed the poverty of a regulative approach dominated by minimal fulfilment of an undemanding body of legislation.

The reports of the mid-1860s do nevertheless provide details of waterworks technology at that time, as well as prepare the way for a discussion of the increasingly sophisticated theoretical debate on the nature and limits of filtration which followed Netten Radcliffe's seminal investigation in 1870. What progress, the companies were asked in 1866, had been made since the last return on engineering and technical processes ten years earlier? Sceptics were disappointed by the sketchy replies that were received. Of the companies drawing their supplies wholly from the Thames, the West Middlesex had constructed two additional filters and now possessed over eight acres of beds in all. But improvement here was more than outweighed by the performances of the Lambeth and Grand Junction companies: the former could only report that new filters were 'under construction' at Thames Ditton while the latter informed the Government that there had been 'no change'.[16] In the same year William Farr, as part of his monumental report on the cholera epidemic of 1866, confronted the companies with a comprehensive set of questions. Once again the answers which he received were less constructive than they might have been—both directors and engineers were clearly deeply suspicious of the motives of a man who, together with Frankland, had brought the East London Company very nearly to its knees. What the responses did reveal, however, were continuing variations over a central aspect of filtration, that is, how frequently the beds were rested and cleaned. The time periods here ranged from monthly (the Chelsea, Grand Junction and Southwark companies) to fortnightly (the Lambeth) to between one and three weeks (the West Middlesex).[17]

There now developed a debate in London between those who took up broadly optimistic and pessimistic attitudes towards the process of filtration. That this controversy was so intimately related to practice and procedure rather than theory is explicable in terms of the still highly empirical state of the emerging 'science of filtration', and the intuition, hardening into an ever-present and haunting fear, that incompetent treatment of water could lead to large-scale mortality from infectious disease and must therefore be rapidly eradicated. Among the optimists Henry Letheby, who succeeded John Simon as medical officer to the City of London in 1855, and made regular analyses of company water on behalf of the Association of Metropolitan Medical Officers of Health, claimed that sand filtration removed significant amounts of dangerous 'organic' matter from public supplies.[18] But he also believed that regular resting of beds and washing of filter sand, as well as the removal of silkweed in the summer months, were prerequisites for a reliable system.[19] Letheby was preoccupied by a problem which would come to dominate the debate over filtration in the later nineteenth century: whether one ought periodically to remove all the sand from a bed and wash it thoroughly to remove all 'impurities', or simply cleanse the surface. Over this he was undecided, but that did not prevent him from taking up a sanguine attitude towards existing technology in the capital—a position which he bolstered with references to epidemiological data showing that other urban areas were more frequently afflicted by water-transmitted typhoid than London. The more pessimistic school, whose position was classically articulated by the Rivers Pollution Commission, held, in the words of Robert Rawlinson in 1866, that 'mechanical filtration separates mechanical impurities: that is to say, that there are no known means of dealing with the effete matter of sewage, when it has once been put into water. . . '.[20] Agreeing, Thomas Orton, the outspoken medical officer for Limehouse, stated that 'filtration through gravel, earth, sand, or animal charcoal' would 'only . . . find it (the water poison) break through these impediments with all its pristine vigour, and become as active as ever in the battle of death'.[21] John Simon was unconvinced that filtration could be relied upon to rid water of anything other than these 'gross mechanical impurities' and he told the Royal Commission of 1868 that he still used a charcoal filter in his own home.[22] But it was Edward Frankland who offered the most articulate and detailed critique of current filtration practice. There was, he argued, a continuing and dangerous tendency for the companies to allow unfiltered water to mix with supplies that had already been subjected to purification.[23] He urged the companies to ask themselves whether intermittent filtration through one foot of sand might not be equally, if not more, effective than the existing practice of relying upon five or eight. And he insisted that current waterworks theory had little if anything to say about the minimum thickness of sand which should be left on the surface of a filter after dirty material had been scraped away. This, according to Frankland, was an exceptionally important point, since it was 'the surface of the sand which, when exposed to the sun, condenses the oxygen out of the air and uses up that oxygen afterwards for oxidising the organic matter'.[24] Here he raised the crucial issue of the 'effective surface' of the filter—a theme which would recur during the next 30 years and which would be refined and reformulated in the light of increasing bacteriological knowledge.

Events in south London, meanwhile, centring on the quality of water delivered by the Southwark and Lambeth companies, appeared to lend support to those who had taken up a sceptical attitude towards filtration in the capital. Widespread popular complaint led to an investigation by Netten Radcliffe, on behalf of the Medical Officer of the Privy Council, which redefined the context within which filtration itself should be assessed and established a comprehensive code to which the companies were urged, albeit without success, to lend their support. The medical officer for St John's and St Olave's, Southwark, had complained that, throughout 1869, water delivered to his area had been similar to 'diluted pea-soup or to a yellow November fog'.[25] A cursory inspection convinced Radcliffe that these observations were sound. What puzzled him, though, was that the Southwark Company drew its raw water from precisely the same intakes as the West Middlesex and Grand Junction concerns and that these two latter enterprises were now beginning to receive extensive commendation for the quality and consistency of their supplies. There was an even more intriguing aspect to the problem— because they were managed by the same engineer, Joseph Quick, both the Southwark and Grand Junction companies made use of very nearly identical filtration strategies.[26] How was it, then, that while one company was providing satisfactory water, the other had become the object of repeated flurries of popular discontent? The general conclusion of Radcliffe's report was that, under conditions of rapidly increasing demand for company water in south London, storage deficiencies had precipitated recurrent overload on the filter beds; and that this overload had, in turn, led to frequent short-term upward movements in the rate of filtration. (The fluctuations had not, however, dramatically affected the long-term *average* rate.)[27] When Radcliffe went on to consider the shortcomings of the Lambeth Company his starting point, as with the Southwark, was that this widely criticised concern was managed by the same engineer as the Chelsea, which was now acknowledged to be delivering a moderately wholesome domestic supply. Filtration processes were very nearly identical in both companies, but the areas devoted to filter beds were not, with the result that the Lambeth was working its plant far too heavily, at an average rate of 55 gallons per square yard per hour, as compared with a figure of 43 for the Chelsea.[28] But the Chelsea Company gained little comfort from Radcliffe's analysis, since he insisted that, during adverse weather conditions, neither concern possessed sufficient plant to ensure the delivery of adequately filtered water.[29] Extreme variability in the rate of filtration, and hence also in the interaction between storage reservoirs and filter beds under adverse climatic conditions, was confirmed by quantitative data which were more reliable than any previous returns (see table 2.1).

**Table 2.1** Filtration rates in gallons per square yard per day for London companies drawing from the Thames (1870). Source: 'Report by J. Netten Radcliffe on the Turbidity of Water Supplied by Certain London Companies', *Twelfth Annual Report of the Medical Officer of the Privy Council:* PP 1870:XXXVIII:Appendix 5: 590.

| | |
|---|---|
| West Middlesex | 235 |
| Southwark and Vauxhall | 341 |
| Grand Junction | 387 |
| Chelsea | 866 |
| Lambeth | 1284 |

Radcliffe claimed that the London companies had failed to regulate themselves or to cooperate with minimal external control. He therefore advocated the adoption of a code of conduct which located the process of filtration within a total waterworks system. This required the keeping of a daily register, which would be open to scrutiny by any duly authorised 'public officer'. Here would be recorded the height and state of the river at the point of intake on any given day, together with the area of filter beds being cleaned or out of use. Information would also be kept on the condition of the filtering surface, the nature and depth of deposits found upon it, and the ease with which water flowed through the filters. Finally, the companies would be charged with monitoring the volume of water delivered between 'midnight and midnight' on every day of the year.[30]

Radcliffe's objectives—that the companies should monitor their activities with greatly increased zeal—were reinforced and reiterated throughout the 1870s. But the context within which discussion of filtration was carried on now underwent significant change. The influential and long-standing Rivers Pollution Commission was ready enough to praise those companies which took steps to improve their procedures but simultaneously, and at times confusingly, issued warnings that filtration could have little more than a minimal effect on a river so overwhelmingly polluted as the Thames. 'The filter', the Commissioners reported in 1876, '. . . supplies to the consumer, in some slight degree, the safety which the wire gauze surrounding his lamp affords to the miner.'[31] 'There is a great difference', they stated elsewhere, 'between this perfect process ['natural filtration'] and the filtration through a couple of feet of sand which is practised by water companies drawing their supplies from polluted rivers: nevertheless evidence is not wanting to show that even this slight treatment is attended with a marked beneficial result; the improvement being due chiefly to the removal of impurities in suspension, but partly also to the oxidation and removal of organic matters in solution.'[32] If this seemed to hold out limited hope for those who championed the Thames rather than alternative non-river sources for the supply of the capital, it was effectively neutralised by the assertion that 'we desire it to be distinctly understood that, although this purification of water polluted by human excrements may reasonably be expected to be some safeguard against the propagation of epidemic diseases, there is not in the form of actual experience a tittle of trustworthy evidence to support such a view'.[33] Reliance on domestic filters was equally dangerous—such mechanisms became easily clogged and were mishandled by 'average servants' with the result that the meagre benefits derived from company filtration were invariably reversed.[34]

Improvements in the 'wire gauze' 'surrounding the lamp' were, however, recorded and commended. That quantitative estimates of rates of filtration were becoming increasingly reliable is supported by the fact that the 'rank order' preferred by the Rivers Pollution Commission was identical to that which had been arrived at by Radcliffe four years earlier. The eight acres of filter beds now operated by the West Middlesex Company enabled it to deliver water deemed to be 'very good'.[35] The Grand Junction, with seven and a half, was still attempting to purify supplies which were in too raw a state and which would have benefited from increased subsidence; filtration at the Southwark and Vauxhall works was judged to be unreliable[36]; and the Chelsea and Lambeth concerns were condemned for continuing to pass 'the greatest volume per hour through a given area of their filter beds', and for delivering turbid water.[37]

And yet there remained an element of contradiction between the belief that sewage-contaminated water was almost certainly irrevocably tainted with the 'poison' of disease and strenuous support for improved levels of filtration. What may in fact be detected between the mid-1870s and the mid-1880s, when increased scientific understanding of the bacteriological mechanisms underlying filtration coincided with a brief period in which the bulk of the water delivered by the London companies satisfied all but the most scrupulous of their customers, is a tension between deeply pessimistic attitudes towards the viability of the 'wire gauze' and a deep-rooted and classically nineteenth-century progressivist commitment to technical innovation. Pessimism was reflected in the observation of A H Hassall, the chemist and naturalist, that 'if the process of filtration is not efficient enough to remove all those more considerable and well known creatures which are named, described and figured in scientific books, it certainly must fail to remove minute cholera germs'.[38] Frankland's reports were equally lacking in reassurance. Only the exemplary West Middlesex Company, he claimed in 1874, possessed sufficiently sophisticated storage and filtration plant to purify Thames water so frequently made turbid under adverse climatic conditions.[39] The Grand Junction and Southwark companies had both failed to respond to repeated criticisms of their treatment procedures.[40] Sudden and unpredictable upward movements in filtration rates were a major problem at the Southwark; but the Lambeth, which consistently recorded an average rate over three and a half times higher than the West Middlesex, generated even greater anxiety.[41] Perhaps, the government auditor reflected in 1876, the real weaknesses were structural and financial: could it really be the case that the companies were incapable of establishing formal capital depreciation funds which would allow them to raise and standardise purification processes?[42]

Within a decade Frankland announced that the overall quality of water consumed by Londoners was unprecedentedly high and that filtration had played a major role in this improvement.[43] What might seem at first sight to have been an inexplicable volte-face was in fact a gradual adjustment to scientific discoveries and to the way in which these were applied, albeit unevenly, to the filtration of river water. 'Filtration', wrote Percy Frankland, Edward's bacteriologist son, 'is no longer a process which should be guided by empirical rules. This most important function . . . is dependent upon principles with which we are intimately acquainted, and which are largely under our control.'[44] This was an overstatement, and there continued to be explicitly 'engineering' debates about the type and thickness of sand which provided the best and most consistent results, as well as complaints that controlled experiments into the effect of changes in filtration rates on delivered water had never been undertaken in this country.[45] Edward Frankland himself vacillated between measured optimism and dark pessimism, and significant numbers of medical men and specialists in public health viewed filtration as a highly fallible 'last defence' against the invasion of sinister and unpredictable germs. 'Supposing', Louis Parkes pondered in 1887, 'an epidemic of cholera or typhoid were to affect the towns on the upper reaches of the Thames and supposing that in every case the filtration should fail in any degree, it was possible that the disease germ might pass into the mains of the London companies and so cause disease. That seemed . . . the great danger of trusting to an artificial system of filtration.'[46] Senior officials at the newly established London County Council were also wary of existing filtration policies. They had a political case to make—municipalisation of water supply became a central progressive demand during the 1890s—but there was also concern that the companies were failing to ensure that 'the conditions of filtration [were] strictly regulated so as to secure uninterruptedly a result which should be precisely defined'.[47]

The Royal Commission of 1893 revealed a truly kaleidoscopic range of attitudes towards filtration and the now more coherent body of theory on which it was based. Some, like William Ogle at the Registrar-General's Office, asserted that a greatly reduced typhoid rate was sufficient to demonstrate that a combination of 'natural aeration', improved sewage systems on the upper river, and filtration had assured the capital of a water supply which was very nearly beyond reproach.[48] Others, though, bacteriologists as well as chemists, were less sanguine. Edward Klein said, and here he had the support of Edward Frankland, that 'however carefully the filtering is done, however much you reduce the number of organisms introduced originally there would still remain a certain proportion that could produce harm'.[49] But on two issues there was something approaching agreement. Firstly, intercompany differentials in filtration rates and policies could no longer be tolerated. 'We cannot shut our eyes', the Commissioners stated, 'to the fact that the provision for these purposes [subsidence and filtration] differs enormously in the different companies, and in some of them is to our mind quite inadequate.'[50] 'What one company has considered it desirable to do', the government water inspector insisted, 'and has done, and has not considered it unreasonable to do, it is fair and reasonable to call on other companies to do.'[51] This was heartening indeed for progressives who championed the creation of a single unifying water authority for the capital: and it marked, also, the end of an epoch in which 'engineering tradition' had repeatedly overridden the recommendations of those who had probed and criticised company policies. The second point of agreement was technical. 'The recommendation', the Commissioners commented, 'which was commonly given in former times, that a filter-bed should be cleansed as often as possible appears to have been a mistake; cleansing, by which the efficient superficial membrane is removed, should only be carried out when the filter has become unduly blocked.'[52] Filters, according to Percy Frankland, should be cleaned neither too often nor too infrequently. 'By keeping your filter beds of moderate age', he explained, 'you can maintain a comparatively rapid process of filtration—efficient filtration—whilst if your beds are allowed to grow too old, probably some other beds will have to be rather over-taxed in order to make up the volume of water which is daily required.'[53] So the minority of engineers who, at mid-century, had favoured filtration, but had also advocated regular 'deep' cleaning of soiled sand, had almost certainly been acting counterproductively. The miner's 'iron mask' of the 1870s was now known, in the 1890s, to be a colloidal surface. Not everyone would have agreed with the exact detail of Sims Woodhead's admonition that '*The sand filters should never be less than three feet thick*, should be disturbed as seldom as possible after they are once in working order, and should never be brought into use until the biological filtering layer has been formed.'[54] But the underlying rationale was now irrefutable: the 'barrier' over which controversy had intermittently raged for nearly 40 years had been shown to be a living organism.

And, yet, still in the mid-1890s and beyond, there were those who distrusted the efficacy of the companies' filtration procedures. The very subtleties and symmetries of bacteriology were themselves depicted as fragile and ephemeral. Biological 'collapse' was argued to be ever-imminent and, if and when it occurred, the city would once again be ravaged, as it had been in the middle of the century, by disease-producing microbes.[55] There was, in fact, no genuine resolution, even among those academics who accepted the new tenets of the 'science of filtration' of that tension between optimism and pessimism which had for so long dominated the 'water question' in London. The validity of this generalisation may be demonstrated by setting the views of two hardy optimists—the eminent James Dewar, and William Crookes, both of them consistent apologists for the water companies—against those of Edward Frankland at the very beginning of the twentieth century. 'Even as late as 1868', Dewar said, 'no person dreamt that sand filtration removed bacteria at all. In fact it was entirely antagonistic to scientific opinion. It is only in recent years that we have found out that what engineers had done antecedently is quite competent to deal with these newer refinements. . . .'[56] Crookes deployed the position developed by Dewar—that since some engineers had been doing the right thing for the wrong reason, the entire armoury of waterworks technology, developed according to 'tradition' rather than 'science', possessed an inherent and timeless validity—to oppose the arguments of sceptics who demanded yet further improvements. Flood water, he insisted, did not pose distinctive problems for the company engineer: and there was still no body of practice or theory which could offer guidance as to how frequently a filter bed ought to be cleaned. 'That', Crookes insisted, evidently either ignorant of or deliberately turning a blind eye towards the powerful corpus of bacteriological information which had been built up during the 1890s, 'is an engineering question we have not gone into.'[57]

Frankland, on the other hand, made use of new insights and analogies both to refute Crookes and to restate the dangers that he had attributed for more than 30 years to the epidemiological potential of brackish floodwater. 'When the Thames is highly charged with microbes', he said, 'the number in the filtered waters supplied by the companies is larger than it is when the Thames water is purer and freer from microbes.'[58] Nor was this all. 'Tradition' and crude 'experience' were inadequate guides to future practice. Intercompany differentials should be analysed in an ever more detailed manner and then eradicated— there was, for example, a clear-cut correlation between small-grained sand as used by the West Middlesex Company and a steady and safe rate of filtration.[59] Why had the other companies refused to acknowledge such unambiguous evidence? Immediately before and after the creation of the Metropolitan Water Board in 1904 there was continuing ambivalence towards filtration and the body of theory on which it was now known to depend. The notorious Maidstone and Lincoln typhoid epidemics in 1897 and 1904 respectively, which resulted in nearly 3000 notifications, took on a deep and frightening significance.[60] 'The Lincoln outbreak', it was said in 1905, 'was only one of a series of cases in which sand filtration had failed to afford security. With regard to the London supply derived from the river it was believed that the sand filtration was not continuously efficient: the influence of the water supply on the prevalence of typhoid fever in the metropolitan area was a matter as to which they had no exact information.'[61]

The persistence, then, of large and moderate-sized epidemics of water-transmitted infection in the early twentieth century cast continuing doubt on the validity of the 'science of filtration'. There were also organisational and engineering problems. By 1908 most towns had adopted slow sand filtration, but many had encountered difficulty in acquiring areas of land large enough for the construction of filter beds. Specialist staff, to operate the plant at the correct speed, were difficult to find. And labour costs for building, maintenance and cleaning could be surprisingly high. Given all this, we should not be surprised to learn that nearly everywhere in Edwardian Britain domestic water was heavily impregnated with peat, iron and other, unknown, discolouring agents.[62]

'In this period', it has been said, 'efficient slow and rapid methods of purification and other processes were perfected.'[63] But such a generalisation does scant justice to the complicated processes and debates which have been described in this chapter. It is legitimate to point to technical improvements and to a growing realisation that filtration would only be effective when combined with a careful selection of river supplies and controlled subsidence. The West Middlesex Company, which had been subjected to savage criticism in the mid-nineteenth century, had clearly learnt many of the lessons which had been passed on to it by investigators of the metropolitan 'water question'. But why, if one concern was capable of introducing technical innovations and more efficient procedures, were the others unwilling or unable to follow suit? The Grand Junction Company appears to have made significant improvements both to its plant and the quality of its water but the other Thames concerns—the Chelsea, Lambeth and the Southwark and Vauxhall—were regularly and, so it would seem, legitimately castigated for inefficiency and technical backwardness.

But conclusions cast exclusively in terms of technological innovation, or the failure to innovate, cannot encompass the full scientific and, beneath that, the ideological and cultural complexity of the theory and practice of water purification in London in the nineteenth century. We have seen that, while some mid-century water engineers championed filtration, others believed it to be little more than a 'sentimental' luxury. During the next 30 years, even though their actions might reveal them to be less than wholly enthusiastic converts, all those professionally concerned with water supply in the capital came to be persuaded that filtration was an essential part of good waterworks practice. What demands further examination, however, are the ways in which, from the 1880s onwards, growing bacteriological knowledge affected specialist and lay attitudes towards traditional methods of water treatment. And here the evidence indicates that, far from encouraging universally positive attitudes towards water derived from the Thames, new scientific theories seem to have led, for many, to increased rather than reduced levels of anxiety. But why? Two possible explanations have already been touched upon. The first, and most obvious, is that any rapidly developing body of knowledge and speculation tends to generate questions and hypotheses which, far from confirming existing accounts of reality, make them more problematic. The second is that the structure and language of bacteriology in the late nineteenth and early twentieth centuries gave rise to metaphorical representations of the physical world which could seem to be both bizarre and threatening.[64] The movement from the 'mechanical' to the 'biological', from the 'visible' to the 'invisible', from 'animaliculae' to 'microbes'—when applied to traditional methods of filtration, each of these images might lessen rather than enhance confidence. If the 'efficient surface' of the filter were really so delicate, surely it could easily be torn or destroyed? And if the 'living biological barrier' broke down, surely millions upon millions of indestructible germs, each carrying deadly infection, would flow into the London mains? When we turn from popular imagery and fear of science in late-nineteenth-century London, and the belief, in particular, that the filters might break down and the capital be afflicted, as Maidstone and Lincoln had been, by water-transmitted infection, to the way in which these matters were viewed by public health specialists, different types of analysis are required. Suspicion of river sources grossly polluted by sewage had its origins in the 1830s and 1840s and was linked to prevailing assumptions about order and disorder, and of the 'natural' and profoundly 'unnatural' in the new urban environment. These ideas received explicitly scientific expression both in the reports of the Rivers Pollution Commission between the mid-1860s and the mid-1870s and Edward Frankland's regular assessments of the Thames for the Registrar-General. Although by the mid-1880s a cataclysmic attitude towards the dangers of consuming water which had been in contact at some period of its 'life' with human faeces had been modified, elements of the earlier pessimism remained latently influential. And, in this sense, bacteriological knowledge could do so much but no more to eradicate so deeply embedded a set of beliefs. The political and social collapse which had been envisaged and feared by the ruling class in the 1840s had not materialised. But the images and taboos associated with it—uncontrolled urban growth and in-migration leading to massive and alien flows of waste into rivers long idealised as fundamental symbols of the healthy and the orderly—could not be eradicated solely by the development of a new corpus of scientific knowledge.

1    H W Dickinson, *The Water Supply of Greater London* (1954), 105 and 120.
2    *SC Metropolis Water Bill* (1851) Q 12 486.
3    Cited in *Report on the Cholera Epidemic of 1866 in England* PP 1867–8: XXXVII:362–5.
4    *SC Metropolis Water Bill*, Q 13 210.
5    *Ibid*, Q 10 896.
6    *Ibid*, Qs 11 898–9.
7    *Ibid*, Qs 3829 and 3842. Evidence of Professor W T Brande.
8    *Ibid*, Q 3746.
9    *Report on the Chemical Quality of the Supply of Water to the Metropolis* PP 1851:XXIII:407.
10   *Report of the Committee for Scientific Enquiries in Relation to the Cholera Epidemic of 1854* PP 1854–5:XXI:47.
11   *General Board of Health: Reports to the Rt Hon William Cowper MP on the Metropolis Water Supply* PP 1856:LII(Pt II):350.
12   *Ibid*, 307.
13   *Ibid*, 286.
14   *Ibid*, 297.
15   *Ibid*, 299.
16   'Copy of a Letter addressed by direction of the Secretary of State of the Home Department to the Metropolitan Water Companies . . . with a Tabular Summary of the Replies of the Companies Thereto' PP 1866: LXVI:672–3.
17   *Cholera Report* (1866) 362–5.
18   *Select Committee on East London Water Bills* PP 1867:IX:Q 2279.
19   *Ibid*, Qs 2280 and 2289.
20   *SC Thames Navigation Bill* (1866) Q 2715.
21   'Special Report by Thomas Orton, MOH, Limehouse, on the Cholera Epidemic of 1866', 16.
22   *RC Water Supply* (1868) Qs 2795–6.
23   *Select Committee on Metropolitan Water (No 2) Bill* PP 1871:XI:Qs 1175 and 1179.
24   *Ibid*, Qs 1340 and 1159.
25   *Twelfth Annual Report of the Medical Officer of the Privy Council:* Appendix 5: 'Report by J. Netten Radcliffe on the Turbidity of Water Supplied by Certain London Companies' PP 1870:XXXVIII:768.
26   *Ibid*, 771.
27   *Ibid*, 774–5 and 779.
28   *Ibid*, 781 and 783.
29   *Ibid*, 784.
30   *Ibid*, 787.
31   *Rivers Pollution Commission: Sixth Report* (1874) 548.
32   *Ibid*, 546.
33   *Ibid*, 551.
34   *Ibid*, 549.
35   *Ibid*, 608.
36   *Ibid*, 610–11.
37   *Ibid*, 548.

38  A H Hassall, *Food: Its Adulterations and the Methods for their Detection* (1876) 44.

39  'Report on the Analysis of the Waters Supplied by the Metropolitan Water Companies during the several months of the Year 1874. By Professor Frankland . . .' PP 1875:XXXI:227.

40  *Frankland's Report*, PP 1877:XXXVII:218.

41  'Metropolitan Water Supply: Report by Lt-Colonel Frank Bolton, Water Examiner under the Metropolis Water Act, 1871' PP 1880:XXVI:281.

42  'Report by Allen Stoneham Esq, on the Audit of the Accounts of the Metropolitan Water Companies during the Year 1875' PP 1876:XXXI:271.

43  *Frankland's Report* PP 1887:XXXVI:370.

44  Percy Frankland, 'On the Filtration of Water for Town Supply', *Trans. San. Inst. GB* **viii** (1886–7) 283.

45  'Report for the Year 1887, by Major-General A. de C. Scott, Water Examiner appointed under the Metropolis Water Act, 1871' PP 1888:XLIX:348.

46  Louis Parkes, *Trans. San. Inst. GB* **ix** (1887–8) 393–4.

47  *Report of the MOH to the London County Council* (1893) Appendix 7, 4.

48  *RC Water Supply Metropolis* (1893) Qs 10 300–16.

49  *Ibid*, Qs 10 989 and 12 613.

50  *Ibid*, 73.

51  *Ibid*, Q 12 972. Evidence of Major-General A de Courcy Scott.

52  *Ibid*, 64.

53  *Ibid*, Q 11 097.

54  *Ibid*, Appendix C70 504.

55  *Report of the MOH, Wandsworth* (1895) 27.

56  *RC Water Supply Metropolis* (1900) Q 21 444.

57  *Ibid*, Qs 21 574, 21 589 and 21 599.

58  *Ibid*, Q 21 979.

59  *Ibid*, Q 28 818.

60  W M Frazer, *A History of English Public Health 1834–1939* (1950) 173 and 295.

61  A G R Foulerton, *J. San. Inst. GB* **xxvi** (1905) 677.

62  William R Smith, 'Modern Methods of Filtration', *J. R. Inst. Public Health* **xvi** (1908) 259.

63  Charles Singer and E A Underwood, *A Short History of Medicine* (second edition, 1962) 231.

64  On metaphorical representations in science see David Bloor *Knowledge and Social Imagery* (1976). On fear of environmental images and realities see Mary Douglas and Aaron Wildavsky, *Risk and Culture: An Essay on the Selection of Technological and Environmental Dangers* (Berkeley, 1982) chapters 6 and 7, and Mary Douglas, *Purity and Danger: An Analysis of Concepts of Pollution and Taboo* (1966) *passim*.

# 3 Chemists and Others

The underlying contention of this chapter, which is devoted to water analysis in London during the second half of the nineteenth century, is that an environment can never be directly and 'naturally' experienced. Rather, ideological preconception and scientific or proto-scientific theory jointly shape the 'ways of seeing' that enable individuals and social groups to articulate and then to seek to reduce anxiety generated by perceived levels of pollution. Little attempt is made, therefore, to evaluate the 'achievement' or to relate the internal history of water analysis during this period. Discussion is directed, rather, towards relations and conflicts between broadly chemical and medico-environmental accounts of river pollution in the capital, to the professionalisation of water analysis, and to the extent to which chemical analysts came directly under the influence of the water companies. The manner in which bacteriological insights in the 1880s and 1890s led to a reformulation rather than a wholesale replacement of existing explanations of the spread of infection via the water 'poison', which had been hypothesised by the more creative and eclectic of the chemists during the preceding 25 years, is also examined. And the ubiquitous themes of 'optimism' and 'pessimism' are related to the intense in-fighting which characterised the profession of water analysis in London between the 1860s and the 1880s.

The historical sociology of scientific knowledge is a difficult and underdeveloped field and it will become clear that, while a credible explanation may have been given of the social roots of a body of chemical theory which was pessimistic in its attitude towards the water 'poison', the optimistic stance is less comprehensively dealt with. But it can be contended that the arguments subscribed to by those who believed that water derived from the Thames in the later nineteenth century was 'beyond reproach' were relatively non-problematic. In this respect, significantly lower death rates in London from infections believed to be transmitted by unsafe water than in nearly any other European capital, appeared to make rumination on the water 'poison' or on 'germs', and the chemical evidence from which their presence might possibly be inferred, largely irrelevant.

'Probably the greater part', a medical man commented in the late 1820s, 'of whatever is communicated to the water from the decomposition of organic matter escapes the action of the tests employed in chemical analysis.' 'Nor can chemical experiment alone determine any thing with regard to salubrity, which is a question that can be solved only by experience.'[1] A decade or so later a surgeon confessed that he doubted the usefulness of the 'eye-test'—the examination of water held up to the light in a glass tube. It was, he said, 'very fallacious', and even though water delivered by the West Middlesex Company might *seem* to be clear, it was in fact almost certainly impregnated with urine.[2] But there were, of course, many water analysts who had full confidence in their procedures and expertise in mid-nineteenth-century London. One of them said that he would stake his reputation on the fact that there was 'no trace of sewage matter in the water at Battersea'.[3] Nor did he know of any 'writer, chemist, physician, or physiologist who has ever adduced a single fact to show injury arising from the use of water containing microscopical animals . . . I believe we would eat nothing and drink nothing if we used the microscope beforehand to settle that point.'[4] Such statements aroused the anger, in particular, of naturalists who prided themselves on their microscopical expertise and who were also sceptical of the unpredictable and often contradictory findings of chemical analysis. Although, as we have already seen, there was at this time no agreed body of theory relating the consumption of dirty water to outbreaks of river-transmitted disease, both medical men and the lay public were convinced that supplies laden with filth and 'animal life' must in some sense be detrimental to good health. To argue, therefore, that the swallowing of amoebae was a harmless activity was *contra natura*, while the assertion that literally everything eaten or drunk in the capital was swarming with alien organisms smacked of the incredible.

At precisely that moment, then, when the quantity, quality and control of the water supply of the capital became the subject of intense debate, the expertise and integrity of water analysts were also called seriously into question. An evaluation of the testimony of chemical consultants, and particularly those employed by the water companies, is a complex undertaking, and can be seen to be so when one considers the varying degrees of credibility that can be attached to the statements of scientists working on research and development for corporations and state-controlled agencies in the later twentieth century. Important elements of the truth are occasionally revealed, but only rarely the whole truth—were it to be otherwise, scientific careers as well as the social and moral legitimacy of employers would be permanently jeopardised. During the nineteenth century conventions governing relations between those responsible for public services, consumers and the government were barely in their infancy. Generalised *laissez-faire*—in this sphere at least[5]—undoubtedly held sway and the notion of accountability, which is now given widespread if often little more than rhetorical emphasis, was wholly undeveloped. To be a company water analyst in the nineteenth century, therefore, involved a primary allegiance to one's employer: and the conflicts of conscience often endured by scientific and technical personnel in our own times were probably only rarely experienced. Although evidence which might reveal the corruption or probity of company chemists is fragmentary and sometimes ambiguous, there is enough of it to enable us to conclude that scientific observation was frequently, and unproblematically, tempered by loyalty to a client. That, certainly, was how Chadwick saw things at mid-century. 'The power of money', he wrote, 'in procuring scientific witnesses is the most disgusting feature. . . . In favour of Thames water as it now is, the Companies obtained the strong certificates of men of science who had denounced it before the Board of Health.'[6] And over the next 30 or 40 years William Farr, Edward Frankland, John Simon and others would express themselves in similar vein.

But it was the weakness and contradictions of analytical techniques which gave greater cause for concern during the 1850s. 'It is the most difficult thing to discover', Board of Health investigators lamented in 1856, 'the precise amount of . . . poisonous ingredients which may produce . . . injury to the system: while, on the other hand, it is impossible to say how small a quantity of such matter might not, under certain circumstances, be injurious.'[7] The insidious nature of apparently clear water and the inability of existing procedures to give adequate guidance on the subject were also deplored. 'The waters from the shallow wells of London, perfectly bright as they are, frequently present, under examination, evidence of impurities derived from innumerable cesspools and sewers with which the metropolis is riddled and traversed: but those impurities may not be detected by the senses.'[8] 'The opinions of chemists', it was reported in the same year, 'are divided as to the manner in which organic matter is capable, under certain conditions, of producing a deleterious effect upon the animal economy.'[9] The most constructive role that water analysis could play, therefore, would be to point to evidence from which it might be possible to infer the presence of genuinely 'poisonous' matter. This constituted a persuasive programme at a time when it was becoming more widely acknowledged that the 'substances which constitute the organic matter of water act injuriously by no means in consequence of being poisonous themselves, but by undergoing those great processes of transformation, called decay and putrefaction, to which all vegetable matter is subject, when no longer under the control of vitality either in plants or animals.'[10] Water analysis, in other words, should become the handmaiden of greatly more subtle forms of environmental investigation. The difficulty, however, was that such methods had as yet been no more than provisionally formulated.

Within a decade and a half water analysis was to be riven by acute theoretical and methodological differences. The 'eye-test' was of no more than limited value within the context of a discipline that had now been committed to instrumental investigation and quantification for more than three quarters of a century. When carried out by experienced practitioners it could yield useful information (and is in fact still used, in a much more sophisticated form, by present day river pollution officers) but this was outweighed by the fact that even the most transparent liquid might be polluted by an invisible 'poison'. Any new procedure, therefore, must either actually identify invisible carriers of disease or, failing that, provide indirect evidence of their presence. Over this issue the major protagonists, W A Miller, Edward Frankland and Frankland's former assistant at Owen's College, Manchester, J A Wanklyn, were in agreement. What divided them was the most effective way of measuring material that could be directly attributed to dangerous organic matter. In 1865 Miller devised the potassium permanganate process which sought to identify the rate at which 'putrefying matter' absorbed oxygen. Two years later, Wanklyn announced a technique which distilled ammonia from the albumen contained in a given sample of water. And in 1868 Frankland described his 'evaporation process' which involved burning a residue with copper oxide and measuring the resulting quantities of carbonic acid and nitrogen.[11]

In the longer term it was the simplicity of Wanklyn's procedure which gave it primacy over both Miller's approach and Frankland's elegant but time-consuming technique. 'Wanklyn's method', it was said in 1887, 'is easily learnt and easily worked, while Frankland's is most difficult: except in the hands of the most expert it may give an error of experiment greater than the total quantity to be measured.'[12] 'The results of the Frankland process', another authority concluded, 'yield no more information as to the dangerous character of the organic pollution than do the results of any other process. The distinction between animal and vegetable organic matter is made out with as much certainty as the ammonia process properly worked, and duly accompanied by other analytical evidence as by a study of organic carbon to organic nitrogen: and although, as regards the Frankland process everybody must unquestionably admire the great ingenuity and skill which have been brought to bear in the development of that truly beautiful analytic method, yet the sanitarian has not much, if anything, to gain by making use of it.'[13] The medical officer for St Marylebone put matters more succinctly. 'It must be apparent', he wrote, 'that a system of analysis so elaborate, although it may afford interesting information to the scientific chemist, cannot fail to mystify and probably alarm the unscientific public.'[14]

Frankland remained committed to his 'beautiful scientific method'. He attacked Henry Letheby for relying upon the 'quite useless' permanganate process for the monthly reports which he supplied to the Association of Metropolitan Medical Officers of Health. He disputed the assertion that 'by boiling with an alkaline solution of potassic permanganate, most nitrogenous organic bodies are decomposed with evolution of ammonia'. Wanklyn's process, Frankland contended, was incapable of measuring either the absolute quantities of total organic matter or of organic nitrogen: and it failed, also, to isolate either the relative amounts of the two categories in different samples of the same source or to distinguish between albuminoid and other nitrous compounds.[15] As for Wanklyn, he responded bravely to Frankland's intimidating onslaughts and had the satisfaction of seeing a majority of analysts adopt his own procedure.[16] If it is correct to interpret intense debate as an indication that an activity has moved from the domain of the proto-scientific into the mainstream of normal science, the Frankland–Wanklyn controversy was of great significance for water analysis both in London and nationally. But there were still many in the later 1860s who were both critical of the efficacy of analysis *per se* and suspicious of the relationship between several leading London consultants and the water companies. 'Chemical action', William Farr reiterated after the cholera of 1866, 'as the chemists on behalf of their wealthy clients tell us, is incessantly going on, and converting impurities into simple elements: so that it is only in some places, or in rare circumstances that the organic waste can reach and injure the people. Still in this mitigated form the risk is too tremendous to be incurred by two millions and half of the people, who require and can obtain an abundance of sweet water.'[17] Frankland, who was to remain deeply critical of the water concerns throughout his career as analyst to the Registrar-General, despised chemists who were willing, as he saw it, to sell their expertise. But—and this was an even more fundamental source of anxiety—he was now also sceptical of the possibility of ever devising a means of tracing the water 'poison' by exclusively chemical means. 'The impossibility', he wrote during the cholera epidemic of 1866, 'of proving either the presence or absence of choleraic and other allied poisons in water, and the uncertainty of all processes for their removal, renders it the more important to guard with the most scrupulous care against the possibility of contamination.'[18] This was precisely the type of statement which played into the hands of those 'pragmatic' public health officials, like Thomas Orton, who were sceptical of nearly every form of explicitly scientific speculation and experiment. How, Orton asked, was it possible that 'one Professor' had declared the East London Company water unusually bad in 1866, while a 'distinguished chemist' had stated that it had never been better?[19] The 'distinguished chemist' was Henry Letheby and he had told a select committee following the cholera epidemic that there was no evidence 'chemical, physical or pathological that the organic matter in the water supplied to the metropolis is any other than the harmless products of vegetable and infusorial growths'.[20] By the end of the decade Letheby was attempting to convince a public meeting of medical men and others that Frankland's profound pessimism—in relation to the possible epidemiological irrelevance of water analysis, the deadly implications of 'previous sewage contamination', and the unacceptibility of river water as a source for public supplies—was wholly without justification.[21]

But other developments, especially in the fields of epidemiology and applied statistics, suggested that there was an undeniable cogency to Frankland's critique of the efficacy of water analysis. The investigations by John Simon and his colleagues at the Medical Office of the Privy Council between 1858 and 1870 had shown that important and decimating infections, particularly typhoid, diarrhoea, dysentery and cholera, were either partly or wholly attributable to contaminated water.[22] Simon himself was not yet fully committed to the 'exclusive' water theory. But this was less important than that detailed observations of topography, social conditions and the course of given infections in a wide range of localities indicated that water was often the major cause of serious outbreaks of disease. And it was precisely in circumstances such as these that the erratic nature of chemical analysis had been most comprehensively revealed. Epidemiology and social statistics, in other words, were now strengthening the position of those who argued that existing techniques for evaluating the quality of water were incapable of warning when a disease might strike: and, when an outbreak had ended, of throwing light on the likely mode of spread.

John Simon articulated this 'sociomedical' assault on the chemists when he told the Royal Commission of 1868 that 'water might be . . . capable of spreading cholera but chemists would be unable to identify the particular contamination which produces that effect'.[23] He commented, also, on the weakness of water analysis—the weakness, in fact, of all the chemical and biological sciences which aspired towards exactness at a time when the germ theory was perceived predominantly in terms of 'poison'—as a means of tracing the origin and mode of transmission of infection. 'A chemist', he said, 'would perhaps report that the water contained "organic matter" but "organic matter" covers an infinite variety of things, and he would have no means that I know of for discriminating the organic matter which is really the ferment, the infectious material of cholera, from a great number of other organic matters.'[24] Taking Simon's arguments a step further, Sir Benjamin Brodie said that 'medical statistics will tell you more about the injurious or non-injurious character of sewage than any analysis would do'.[25] But, he went on, medical statistics were just as flawed, in their own way, as water analysis. Information on mortality and morbidity might certainly 'elicit relations between cause and effect', but only exceptionally would such data allow even a rough-and-ready prediction of the outbreak of an epidemic. And if it were in fact the case that a water supply was suspected of being contaminated it was self-evident that 'you could not poison people for the sake of trying the water'.[26] Water analysts may have failed to indicate when and why a supply was safe or to develop processes for identifying the water 'poison' which some chemists, like their medical counterparts, now believed to be a major precipitant of disease. But the overall objective—to devise a reliable indicator of when and in what 'strength' a poison *might* be present—was an unexceptionable one. Simon took up and elaborated Brodie's notion of 'experimentation'. 'Water into which sewage has been discharged', he insisted, '. . . is an experiment on the health of the population, and I do not think that the experiment ought to be tried. Moreover, as a mere matter of taste, people would rather not drink water into which sewage has been discharged and I think that that in itself deserves consideration.'[27] Where Brodie had used the term figuratively to show up the differences between the medical and chemical styles of investigation, Simon quite simply accused the companies of knowingly deriving their supplies from grossly polluted sources, of 'experimenting with human life'. This was the context within which investigatory bodies, like the Royal Commission of 1868, looked to men of science for guidance on questions affecting public water systems, and they were profoundly disappointed with what they encountered. 'We have found', the Commission stated, 'not only that opinions are divided upon it [the quality of the Thames] but that the elements which enter into its determination are of a very subtle character, and by no means admit of the satisfactory kind of treatment which we are in the habit of expecting from the modern advanced state of physical science.'[28] If external criticism of chemical analysis was severe at this time, the profession was also still deeply internally divided, both in London and nationally. Letheby continued to attack Frankland, characterising the latter's assessments of the Thames as 'sensational' and 'calculated to mislead public opinion'. Frank Bolton was equally critical. A detailed account of the river, in terms of organic pollution; the climatic conditions influencing the quality of water taken into storage reservoirs; rates of subsidence and filtration; and descriptive evaluations of delivered supplies as 'turbid', 'cloudy' or 'unfit for consumption'—all these should be replaced by an index which would simply impress upon the companies 'a minimum of solids above which they ought not to go'.[29] Bolton also held that it 'is to be regretted that such terms as "living organisms" and "moving organisms" have been used so frequently and indefinitely'. 'It is well known', he wrote, 'that it is impossible altogether to get rid of the simplest forms of vegetable and animal life . . . even by the most perfect filtration.'[30] But Frankland was as ready to stand firm against Bolton as he was to harangue the companies and their chemists.

When he attempted to summarise the state of water analysis for a non-specialist readership in the mid-1870s, a period when commitment to an imprecise form of the germ theory combined uneasily with a conviction that chemical examination might well be useless in the face of the 'invisible poison', Arthur Hassall told his audience that very few practitioners now actually equated 'organic matter' with the *materies morbi*. Rather, when taken in conjunction with other contextual factors, the quantity of disease-producing material believed to be present was intended to give no more than an indication of 'the degree of liability of any water to become contaminated with the special poison of cholera, typhoid fever, or other disease'.[31] It was now also generally agreed that even 'the purest distilled water' might disseminate infection if it were to come into contact with the infectious or contagious matter of typhoid fever'.[32] In terms of the presentation of results, Hassall complained that there was still too widespread a tendency for analysts to coin popular, 'short-hand' descriptions of the quality of a given sample. 'At present', he wrote, 'the purity or impurity of a water is expressed by some such terms as the following—moderately good, good, very good; or, rather bad, bad, very bad; no two persons in using these expressions meaning exactly the same thing, having no rule or standard to guide them: thus a water which to one chemist would be good, to another would be bad.'[33] This was a serious charge but one which could still be applied to each of the analysts and groups of analysts who were active and influential in the capital in the mid-1870s. Different procedures were being deployed and incommensurable quantitative results derived from them: and the translation from the 'quantitative' to the 'qualitative' involved precisely those discrepancies between 'experts' which were deplored by medical men and derided by the general public.

Controversy over the best chemical method of evaluating the safety or otherwise of drinking water rumbled on into the 1880s and was undoubtedly exacerbated by Frankland's extended occupancy of the prestigious and influential position of quasi-official analyst to the Registrar-General. The speed and extent to which the Thames could purge itself of sewage by means of 'self-aeration' and the suitability of rivers as long-term sources for public supply continued to be areas of disagreement. Frank Bolton, in particular, was exasperated by the contrary advice that he received from rival groups of specialists. 'Here Dr Frankland tells us', he said in 1880, 'that the supply of the Kent Company would be a priceless boon, and at once would confer upon it an absolute immunity from epidemics of cholera.' Charles Meymott Tidy, on the other hand, professor of chemistry and forensic medicine at the London Hospital, insisted that the medical and epidemiological record completely nullified such a claim.[34] Tidy told a meeting of the Chemical Society in 1881 that he, Sir William Crookes and William Odling, professor of chemistry at Oxford, were of the 'opinion that the filtered water of the Thames and Lea is unimpeachable in respect of its wholesomeness and suitability for town supply'.[35] What was left unsaid was that each of these experts had at one time or another worked for a water company—and had this information been given wider currency the claim that the death rate in the districts supplied with deep well water by the Kent Company was the highest in London would have been subjected to more intensive scrutiny.[36] The result of this phase of a continuing controversy was that the Thames was not abandoned as a source for public supply. But the debate itself, like its predecessors, had been heavily influenced by social interest and personal rancour. The profession of water analysis in London had been shaped to a significant degree by problems posed by the Thames and this in itself strengthened opposition to alternative sources. Powerful members of the 'pro-Thames' lobby were also deeply involved professionally with companies which drew their supplies from the river and had a clear-cut economic interest in continuing to do so. The Franklandite 'anti-Thames' faction, on the other hand, remained deeply influenced by its preoccupation with 'previous sewage contamination' and the alleged unreliability of nearly every type of river water. (It was, of course, objectively undeniable that very large numbers of Londoners had perished between about 1830 and 1870 as a direct result of drinking water heavily impregnated with human waste, but this was not an argument which could be so readily sustained by the early 1880s.)

Commitment to specific analytic techniques, then, and the conclusions as to the most reliable source of urban water supply systems to which they gave rise, had for many years been heavily influenced by non-scientific considerations. When, at a meeting of the Association of

Metropolitan Medical Officers of Health in 1869, Henry Letheby, supported by Wanklyn and Odling, had attacked Frankland's conception of 'previous sewage contamination', he immediately found himself having to repudiate 'the imputation of having any communication whatever with the Water Companies'.[37] In 1877, Meymott Tidy, who had succeeded Letheby as analyst for the Association on the latter's death in 1876, made scathing reference to the Rivers Pollution Commission and what he believed to be their total ignorance of the medical and epidemiological aspects of the 'water question'.[38] But it was not until 1880 that the storm finally broke. 'When', it was written some years later, 'the analysis of the Metropolitan water supply was entrusted to Dr Tidy in 1876, there was an understanding that he should have nothing to do with the Water Companies. At the Annual General meeting of this year [1880] this question was brought up, and on a communication being addressed to Dr Tidy, he resigned his membership of the Society. Thereupon the Council recommended the discontinuance of the monthly analysis—a work which had been prosecuted uninterruptedly for a quarter of a century.'[39] A year later Tidy was renominated for membership of the Association but failed to gain a majority.[40] The documentation of this *cause célèbre*, which had its origins in the long-standing animosity between Letheby and Frankland as well as in Letheby and Tidy's unsuccessful attempt to serve two masters, the Association and the water companies, is scanty. What may be concluded, however, from even the barest outline, is that chemical analysis in London between the 1860s and the 1880s was frequently influenced by considerations which were wholly non-scientific in character.

By the early 1880s a degree of institutional stability as well as a more measured, and less highly personalised, scientific discourse had been established. The implications, in particular, for water analysis of new bacteriological insights were now seriously pondered by academic specialists and by those who had only limited contact with the metropolitan scientific élite. 'It is not possible', wrote the medical officer for Hackney in 1882, 'in the present state of chemical and microscopical knowledge to say that a given sample of water is absolutely harmless because the *state* in which organic matter exists in water is more important than the absolute amount.'[41] Louis Parkes, echoing George Buchanan, argued that 'the chemist can tell us of impurity and hazard, but not of purity and safety'.[42] And within a decade Edward Frankland himself would be deploying bacteriological knowledge and the language of bacteriology to redefine his major preoccupation in relation to public supply systems—the practical steps which could and must be taken to pre-empt the possibility of random outbreaks of water-transmitted disease. 'You cannot', he told the Royal Commission of 1893–4, 'detect the germs by chemical means . . . they are too minute in quantity and weight to affect the results of chemical analysis.'[43] Further and more rigorous research must be immediately undertaken.

Edward Klein was less patient, and his comments at this time may be taken as an extension and summary of the critique which had now been levelled at water analysis, first by medical men and then by bacteriologists, for more than 30 years. Chemical analysis, he claimed, 'is not only not conclusive in the eyes of the sanitarians: it has not got that value which it apparently has in the eyes of many eminent chemists, that they can test for organic impurity, and this water is less pure because it has less than a certain amount of organic matter. Water which may be very much purer in that sense may be charged with infection.'[44] This was an old story but it was now being recounted in a quite novel way. Other long-standing criticisms of chemical analysis were also reformulated. Traditional examination, it was stated during a debate on the resurgence of typhoid in 1898, could only rarely serve as a general predictor of epidemics. 'As a rule water-borne outbreaks of typhoid fever occur with explosive violence; so that most of the victims have imbibed the poison before the alarm conveyed by means of analysis can possibly save them from infection.'[45] This, again, was structurally similar to the critique levelled by Simon, Radcliffe and others in the mid-1860s.

Following the bacteriological researches of Edward Klein and Alexander Houston during the 1890s, the processes whereby water which was 'clear to the eye' might nevertheless transmit large-scale outbreaks of infection began to be explained in more rigorous terms.[46] The elements for such an explanation had, however, long been available and had been stated and restated in a wide range of forums by medical and scientific men in the capital for more than a quarter of a century. In this sense, Frankland's classic dictum in 1893 that the 'water question has, in fact, passed from the domain of chemistry into that of biology'[47] had been prefigured by his own anguished references to the water 'poison' in 1866, when he had suddenly realised that he might be pursuing a phantom and seeking, fruitlessly, to predict the unpredictable.

We have seen in earlier chapters how the problem of pollution of the Thames came to be defined during the nineteenth century and how that definition was related to and partially determined by sociopolitical processes. But what has not yet received sufficient emphasis is that an environment is never non-problematically perceived by those who feel anxiety about it. That this is in fact the case has been briefly illustrated in relation to the rival ideologies which reformers and medical men brought to their evaluation of the external world during the period under consideration, and, in this chapter, by the recurring complaints on the part of analytic chemists that nature was treacherous, ambiguous and impenetrable: the apparently innocuous could be potentially deadly and the seemingly filthy less inimical to good health than commonsense might seem to suggest. It was, then, choice of scientific theory or culturally available world view which determined one's estimate of the safety or danger of a given aspect of the environment. According to this interpretative framework, a theory of pollution which postulates an individual appalled by the desecration of an unmediated nature is replaced by that of social groups of individuals possessing expectations about how nature and society should be ideally ordered and organised: and how those with specialist knowledge, or in the case of water analysis in the earlier nineteenth century, proto-specialised knowledge, actually evaluate specific instances of deterioration and danger.[48] For much of the period discussed here there was no 'science of the environment' but chemical analysis was nevertheless characterised by deployment of discrete, though often disputed, procedures which gave it a limited degree of professional and scientific credibility in the crucial transitional years between about 1860 and 1890. The role of quantification was exceptionally important, for it served both to give the appearance of growing control and to rationalise the failure of the social system as a whole to ameliorate the dominant causes of differential rates of mortality from epidemic disease—a grossly unequal distribution of wealth and income. Scientific procedures interposed themselves between man and nature: they constituted the framework for the emergence of a cluster of new environmental professions (including water analysis) and they tended, invariably, to legitimate the *status quo*.

As for those engaged in water analysis we have seen that they had no option but to be collectively committed to their enterprise. There was internal schism as well as scathing external criticism by medical men, epidemiologists and those concerned with the collection and inter-pretation of demographic data. Denunciations of the relationship between eminent consultants and the water companies were also commonplace but it is as well to remember in this respect that notions of loyalty and corruption were quite different in the nineteenth as com-pared with the late twentieth century and that only a minority of critics—Chadwick, Simon, Frankland, Farr—ever broke even partially out of a discourse dominated by the hegemony of unfettered capitalism and the non-public domain.[49] The apparent inability of chemical analysis to cope either theoretically or practically with the implications of the 'poison' or early germ theory of disease might have been wholly debilitating. But a minority within the profession were ready to appropriate whatever they could from proto-bacteriology and with-stand sceptical interrogation by official investigators during the later nineteenth century. If procedural disputes between 1860 and 1880 had a double-edged effect, involving the consolidation of normal science as well as stimulating acutely embarrassing external attack, the emergence of bacteriology from the 1880s onwards was unambiguously threaten-ing. One must, however, avoid an over-linear or Whiggish explanation of this intensely complicated period in the history of the biological and chemical sciences. The theoretical confusion of the 1860s and 1870s, the classic epoch of the 'poison' theory, had led to a generalised loss of disciplinary autonomy, with medical men borrowing whatever seemed relevant from chemistry and a minority of chemists, as we have seen, seeking an accommodation with epidemiologists and social statisticians. The final transition from the 'chemical to the biological domain' in the late 1880s and early 1890s was, therefore, less traumatic than it might have been; and what emerges from an examination of the period between 1885 and 1900 is that academic background—whether in chemistry, bacteriology, social statistics or public health—was less important than a general appreciation of the relationships between the specificity of disease, the persistence of water-transmitted infection and the most effective means of protecting public water supplies.

It was in London, the scientific as well as the political and cultural capital, that dominant notions of river pollution and latterly of 'environment' were forged and formulated in the nineteenth century.[50] The social theory of pollution and the explanation of the emergence of specific pollution problems at particular times which have been propounded in this section are not, and should not be taken as being, comprehensive. Too great an emphasis has, in particular, been placed on professional science and arcane scientific knowledge, and not enough, for reasons which are primarily evidential, on non-specialist environmental ideologies.[51] What should have been established, though, is that the threat of endemic instability in the 1830s and 1840s had a decisive effect on the formation of pessimistic middle and upper class notions of pollution and environment, and on perceived inter-actions and equations between the natural and the social orders.

1    *RC Water Supply Metropolis* (1828) 144.
2    *SC(HL) Supply of Water to the Metropolis* (1840) Q 668. Evidence of William Clapp.
3    *SC Metropolis Water Bill* (1851) Q 12 284. Evidence of A S Taylor.
4    *Ibid*, Q 12 200.
5    Although the position is clear-cut in relation to public water supply

historians have differed over the respective roles of the state and of unfettered capitalistic individualism in many other spheres. The position is well surveyed by Arthur J Taylor, *Laissez-Faire and State Intervention in Nineteenth Century Britain* (1972).

6    Edwin Chadwick to J T Delane, 20 June 1851 in R A Lewis, *Edwin Chadwick and the Public Health Movement 1832–1854* (1952), 274.
7    *Report on the Work of the Metropolitan Water Companies* PP 1856:LII:345.
8    *Ibid.*
9    *Report on the Quality of the Supply of Water to the Metropolis* PP 1856:LII:257.
10   *Ibid.*
11   Hamlin, *Bull. Hist. Med.* **56** (1982) 63–5. On Frankland see also J R Partington, *A History of Chemistry* vol 4 (1964) 500–1.
12   L Parkes, *Trans. San. Inst. GB* **ix** (1887–8) 380.
13   Charles E Cassall and B A Whitelegge, 'Remarks on the Examination of Water for Sanitary Purposes', *Trans. Soc. Med. Off. Health* (1883–4) 65.
14   *Report of the MOH: St Marylebone* (1868) 31.
15   E Frankland, 'On Some Points in the Analysis of Potable Water', *J. Chem. Soc.* **xxviii** (1876) 347–51.
16   See Wanklyn's comments to *RC Water Supply* (1868) Q 5418–22.
17   *Cholera Report* (1866) 293. Farr was quoting *verbatim* from *Report on the Mortality of Cholera in England 1848–49* (1852), lxxvi–lxxvii.
18   Frankland to the Registrar-General in *Cholera Report* (1866) 227.
19   'Special Report by Thomas Orton, MOH, Limehouse, on the Cholera Epidemic of 1866', 10.
20   *SC East London Water Bills* (1867) 363.
21   R Dudfield, 'History of the Society of Medical Officers of Health', *Public Health*, Jubilee Number 1906, 54–5.
22   R Lambert, *Sir John Simon 1816–1904 and English Social Administration* (1963), part IV. See, also, John M Eyler, 'The Conversion of Angus Smith: the Changing Role of Chemistry and Biology in Sanitary Science 1850–1880', *Bull. Hist. Med.* **54** (1980) 216–34.
23   *RC Water Supply* (1868) Q 2754.
24   *Ibid*, Q 2763.
25   *Ibid*, Q 6991.
26   *Ibid*, Q 7043.
27   *Ibid*, Q 7141.
28   *Ibid*, lix.
29   *Sanitary Report: Paddington* (1871–2) 34–5.
30   Quoted in A H Hassall, *Food: Its Adulterations and the Methods for their Detection* (1876) 48.
31   *Ibid*, 61.
32   *Ibid*, 32.
33   *Ibid*, 57.
34   *SC London Water Supply* (1880) Q 2082.
35   *Report of the MOH: Kensington* (1881) 124.
36   *SC London Water Supply* (1880) Q 1393, George Sclater-Booth to E J Smith; and Q 2083, evidence of Frank Bolton. See, also, on this controversy *Report of the MOH: Lambeth* (1881) 110.

37    Association of Metropolitan Medical Officers of Health, *Minutes*, 1 May 1869.
38    C Meymott Tidy, 'On the Quality and Quantity of the Water supplied to London during the year 1877', Association of Metropolitan Medical Officers of Health, *Secretaries' Reports* (1878–9) 26–8. But Tidy appears to have respected the scientific ingenuity of Frankland's method of water analysis.
39    Association of Metropolitan Medical Officers of Health, *Minutes*, 15 October and 19 November 1880.
40    Dudfield, *Public Health,* Jubilee Number 1906, 102.
41    *Report of the MOH: Hackney* (1883) 20.
42    Parkes, *Trans. San. Inst. GB* **ix** (1887–8) 383.
43    *RC Water Supply Metropolis* (1893) Q 4604.
44    *Ibid*, Q 11 058.
45    Christopher Childs, 'Water-Borne Typhoid Fever', *J. San. Inst. GB* **xix** (1898) 246–7.
46    W M Frazer, *A History of English Public Health* (1950) 176.
47    *RC Metropolitan Water Supply* (1893) 60.
48    Douglas and Wildavsky, *Risk and Culture: An Essay on the Selection of Technological and Environmental Dangers* (Berkeley, 1982) chapter 3.
49    The 'escape', as will be seen in Chapter 8, was often no more than temporary.
50    The role of the 'X' Club, of which Frankland was a member, was important. See Roy M MacLeod, 'The X-Club: A Social Network of Science in Late Victorian England', *Not. Rec. R. Soc.* **24** (1970) 305–22 and J Vernon Jensen, 'The X Club: Fraternity of Victorian Scientists', *Br. J. Hist. Sci.* **5** (1970–1) 63–72.
51    There are tantalising glimpses of the attitudes of working-class women towards the pollution and use of water during this period but the great bulk of available testimony is that of upper middle-class and middle-class men.

# Part II

# Demographic Experiences

One result of this epidemic was to demonstrate, at the cost of thousands of lives, that the system of private water companies supplying the community with this necessity of life was absolutely opposed to the interests of the community.

Henry Jephson on the cholera of 1866 in *The Sanitary Evolution of London* (1907), 192.

# 4 Cholera

The four cholera epidemics (1831–2, 1848–9, 1853–4 and 1866) which afflicted Britain during the nineteenth century brought sudden and traumatic destruction of human life. Epidemic mortality and case fatality, which reached as much as 40 per cent, were exceptionally high, while therapy was no more advanced in 1870 than it had been in 1832.[1] In terms of symptoms, cholera was a terrifying disease. The brief period between contracting the infection, being wracked by violent stomach spasms and diarrhoea, and finally succumbing to dehydration or shock was agonising and only powerful narcotic drugs, which were unavailable to the great mass of ordinary people, might bring respite.[2] When measured against the other major infections of the nineteenth century cholera was also socially traumatic. Once a community had finally been compelled to admit that significant numbers of its citizens had been stricken by the disease, a degree of disruption could be expected to ensue. At the peak of an epidemic the shaky agencies of Victorian local self-government were shown to be inadequate and both general economic activity and family earnings severely affected.[3] Rumour, particularly during the first epidemic of 1832, that the ruling classes had devised a secret Malthusian weapon with which to thin out the over-populous poor, led to further instability.[4]

It is hardly surprising, therefore, that cholera became enshrined as the 'ultimate' fever of Victorian Britain. Because the disease tended to strike most savagely, though not exclusively, in the most poverty-stricken districts of towns, the need for social discipline ensured that it received maximum attention from the Government and a largely help-less medical profession. But precisely because it sometimes spread to middle- and upper-class areas, cholera was all the more deeply feared. It was this element of unpredictability, controverting, so it seemed, both miasmatic orthodoxy which located the origins of disease in the 'epidemic atmosphere' and related sociological explanations which

69

made much of the dissipated and intemperate habits of the poor, that did much to encourage social and scientific reformers to look to the environment and particularly, as we shall see, to unsafe water, as a possible transmitting medium. During this process of investigation and discovery London, the rivers of London—the Lea as well as the Thames—and London's water supply came to be seen as ideal social and epidemiological testing grounds.

Fear of cholera, rather than the actual presence of the disease, was manipulated by reformers to cajole backward sanitary authorities into taking action to reduce the death rate from infections which, year in, year out, were far more destructive than the pandemic itself. Since the general death rate (not to mention infant mortality) declined less rapidly than interventionists like John Simon and Netten Radcliffe believed it should have done, cholera was deployed in this way beyond the final epidemic of 1866. Throughout the period between 1840 and 1870 'backward' districts were held to have inhibited further improvement to national levels of health. And yet undrained and socially insulated towns and villages, which had fortuitously avoided high mortality from cholera, often reacted to official exhortation with obstinate inactivity. Whatever one might do, the élites of such places claimed, cholera and typhus would always return to the same, notorious 'fever haunts'. In such areas—and in the capital they were to be found in the City, immediately to the south of the river and in the East End—poverty, overcrowding, environmental squalor and entrenched localism militated against radical measures.

This chapter is concerned with the quantitative parameters of cholera in London during the epidemics of 1848–9, 1853–4 and 1866, the extent to which these outbreaks were triggered by river pollution and unsafe drinking water, and the degree to which contemporary reformers and medical men perceived dirty water as a determinant of the infection. This last topic is related to a discussion of the interactions between ideas about pollution and that most resilient of nineteenth-century medical paradigms, the miasmatic theory of disease. It will be shown that the final experience of cholera in London in 1866 played a crucial long-term role in undermining existing notions about how the disease was transmitted, and prepared the way, at least among a minority of public health specialists, for acceptance and dissemination of the germ theory of disease. The primary focus is on the period between about 1840 and 1870—a time when many informed contemporaries believed that cholera and pollution, though they could not be linked in a rigorously causal manner, might well destroy both the magnificence and the squalor of the new urban civilisation.

The population and location of the districts discussed in Part II are shown in table 4.1 and the map opposite.

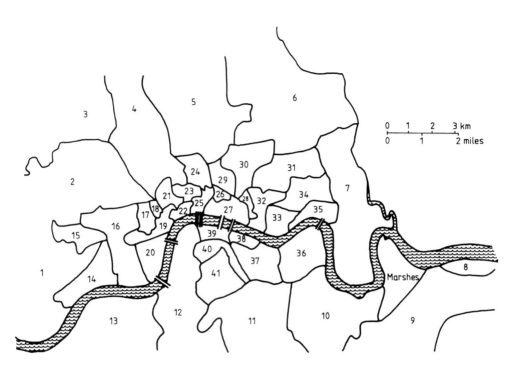

London Registration Districts in 1850. For the sake of clarity individual numbers have been given to parts of districts geographically separated from other parts but bearing the same name.

| | | |
|---|---|---|
| 1 Kensington | 15 Westminster | 28 East London |
| 2 Marylebone | 16 St George, | 29 St Luke |
| 3 Hampstead |    Hanover Square | 30 Shoreditch |
| 4 St Pancras | 17 St James, Westminster | 31 Bethnal Green |
| 5 Islington | 18 Strand | 32 Whitechapel |
| 6 Hackney | 19 St Martin-in-the-Fields | 33 St George-in-the-East |
| 7 Poplar | 20 Westminster | 34 Mile End Old Town |
| 8 Woolwich | 21 St Giles | 35 Stepney |
| 9 Lewisham | 22 Strand | 36 Rotherhithe |
| 10 Greenwich | 23 Holborn | 37 Bermondsey |
| 11 Camberwell | 24 Clerkenwell | 38 St Olave, Southwark |
| 12 Lambeth | 25 West London | 39 St Saviour, Southwark |
| 13 Wandsworth | 26 East London | 40 St George, Southwark |
| 14 Chelsea | 27 London City | 41 Newington |

**Table 4.1** The population of London Registration Districts 1851–1911. Source: *Annual Reports* and *Supplements* of the Registrar-General. Exceptionally rapid rates of growth are frequently accounted for by boundary reorganisations. The overall nature of such changes can be understood by reading the table in conjunction with the map on p. 71.

| | 1851 | 1861 | 1871 | 1881 | 1891 | 1901 | 1911 |
|---|---|---|---|---|---|---|---|
| Kensington | 120 004 | 185 950 | 283 158 | 163 151 | 166 308 | 176 628 | 174 759 |
| Paddington | — | — | — | 107 218 | 117 846 | 143 976 | 143 040 |
| Chelsea | 56 538 | 63 439 | 71 089 | 88 128 | 96 253 | 73 842 | 70 745 |
| Fulham | — | — | — | 114 839 | 118 878 | 249 528 | 261 475 |
| St George, Hanover Square | 73 230 | 87 771 | 155 936 | 149 748 | 134 138 | 128 256 | 123 043 |
| Westminster | 65 609 | 68 203 | 51 181 | 46 549 | 37 312 | 33 081 | 29 520 |
| St Martin-in-the-Fields | 24 640 | 22 689 | — | — | — | — | — |
| St James, Westminster | 34 406 | 35 236 | — | — | — | — | — |
| Marylebone | 157 696 | 161 680 | 159 254 | 154 910 | 142 404 | 133 301 | 126 749 |
| Hampstead | 11 986 | 19 106 | 32 281 | 45 452 | 68 416 | 81 942 | 82 402 |
| St Pancras | 166 956 | 198 788 | 221 465 | 236 258 | 234 379 | 235 317 | 227 626 |
| Islington | 95 329 | 155 341 | 213 778 | 282 865 | 319 143 | 334 991 | 331 326 |
| Hackney | 58 429 | 83 295 | 124 951 | 186 462 | 229 542 | 270 519 | 272 061 |
| St Giles | 54 214 | 54 076 | 53 556 | 45 382 | 39 782 | 31 436 | 29 900 |
| Strand | 44 460 | 42 979 | 41 339 | 33 582 | 25 516 | 21 674 | 20 115 |
| Holborn | 46 621 | 44 862 | 163 491 | 151 835 | 141 920 | 129 432 | 120 956 |
| Clerkenwell | 64 778 | 65 681 | — | — | — | — | — |

| District | | | | | | | |
|---|---|---|---|---|---|---|---|
| St Luke | 54 055 | 57 073 | — | — | — | — | — |
| City | 129 128 | 113 387 | 75 983 | 51 439 | 38 320 | 26 932 | 23 673 |
| Shoreditch | 109 257 | 129 364 | 127 164 | 126 591 | 124 009 | 118 637 | 115 731 |
| Bethnal Green | 90 193 | 105 101 | 120 104 | 126 961 | 129 132 | 129 680 | 128 959 |
| Whitechapel | 79 759 | 78 970 | 76 573 | 71 363 | 74 462 | 78 768 | 73 681 |
| St George-in-the-East | 48 376 | 48 891 | 48 052 | 47 157 | 45 795 | 49 068 | 48 146 |
| Stepney and Mile End Old Town | 110 775 | 129 636 | 150 842 | 164 156 | 164 968 | 170 754 | 168 110 |
| Poplar | 47 162 | 79 196 | 116 376 | 156 510 | 166 748 | 168 822 | 165 814 |
| St Saviour, Southwark | 35 731 | 36 170 | 154 049 | 195 164 | 202 693 | 206 180 | 199 545 |
| St Olave, Southwark | 19 375 | 19 056 | 122 398 | 134 632 | 136 660 | 130 760 | 128 311 |
| Bermondsey | 48 128 | 58 355 | — | — | — | — | — |
| St George, Southwark | 51 824 | 55 510 | — | — | — | — | — |
| Newington | 64 816 | 82 220 | — | — | — | — | — |
| Lambeth | 139 325 | 162 044 | 208 342 | 253 699 | 275 203 | 301 895 | 301 604 |
| Wandsworth | 50 764 | 70 403 | 125 060 | 210 434 | 307 500 | 400 829 | 436 285 |
| Camberwell | 54 667 | 71 488 | 111 306 | 186 593 | 235 344 | 259 339 | 260 223 |
| Rotherhithe | 17 805 | 24 502 | — | — | — | — | — |
| Greenwich | 99 365 | 127 670 | 100 600 | 131 233 | 165 413 | 185 034 | 185 574 |
| Lewisham | 34 835 | 65 757 | 51 557 | 73 327 | 94 335 | 134 721 | 152 746 |
| Woolwich | — | — | 73 380 | 80 834 | 107 324 | 131 086 | 129 511 |

Despite its exceptional virulence, cholera accounted for far fewer deaths (approximately 40 000 in London during the nineteenth century) than probably any other epidemic infection.[5] Mortality from the disease in 1848–9, 1853–4 and 1866 is set out in figure 4.1, which shows that there was a significant differential between London and the rest of England and Wales during each of the epidemics and markedly so in 1853–4. Contrasts of this kind are, however, less revealing than a juxtaposition of metropolitan experience against that of other severely afflicted areas. A comparison of this type is presented in figure 4.2, in which mortality in London as a whole, as well as in the single most seriously affected district within London during each of the outbreaks, is set against mortality in the three most heavily affected urban regions in the rest of the country.

During each of the epidemics under discussion, but particularly during the outbreak of 1853–4, levels of mortality in the capital were significantly influenced by high mortality within specific districts or clusters of districts and this is shown in figure 4.3. Thus, areas to the west and the suburban north-west suffered substantially lower levels of mortality than London as a whole in each of the onslaughts: and some districts, notably Hampstead, Marylebone and Islington, experienced lower death rates than the rest of England and Wales. But a further grouping of districts, including the very poor communities immediately to the south of the river and in the East End, were all traumatically affected by cholera in 1848–9 and 1853–4. Two epidemiologically comparable, though in terms of scale dissimilar, metropolitan outbreaks should also be noted: the vicious eruption, associated with polluted well water in St James, Westminster in 1854; and the major epidemic which afflicted the East End districts of Bethnal Green, Whitechapel, St George-in-the-East, Stepney, Mile End and Poplar in 1866.

When one moves from the quantitative parameters of the disease during this period to an analysis of the ways in which it may have been spread, one approach, first adopted more than a century ago by the Rivers Pollution Commission, is to set aggregate mortality against moderately accurate, though in the final analysis still highly generalised, indicators of water pollution. The Commissioners' conclusions are shown in table 4.2. This method is, however, open to the charge that it evades the issue of transmission by employing mortality from cholera as an index of river pollution, rather than isolating that proportion of deaths from the infection during a given outbreak which should be attributed to the consumption of contaminated water. Polluted water played a major role in each of the epidemics with which we are concerned, but other modes of infection, common to nearly every disease transmitted via the faecal–oral route, were also implicated.

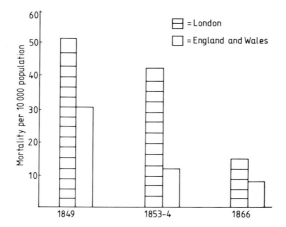

**Figure 4.1** Cholera mortality per 10 000 population during three epidemics. Source: R Thorne Thorne, *The Progress of Preventive Medicine during the Victorian Era* (1888) 59.

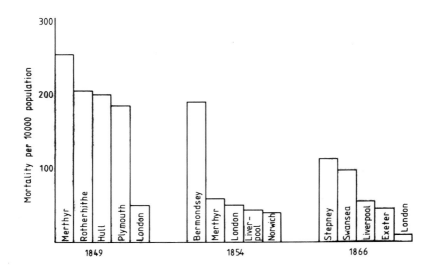

**Figure 4.2** Mortality from cholera per 10 000 population in three epidemics: comparative urban areas. Sources: *Annual Reports* of the Registrar-General and *Report on the Cholera Epidemic of 1866 in England* PP 1867–8:XXXVII:120.

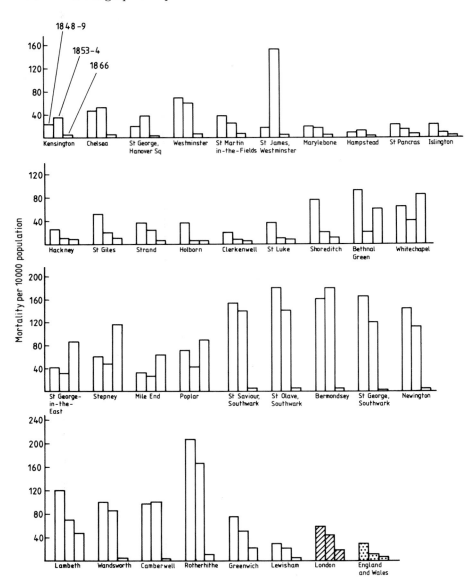

**Figure 4.3** Cholera mortality per 10 000 population in three epidemics: London registration districts. Sources: *Ninth Report of the Medical Officer of the Privy Council* PP 1867:XXXVII:339; *Report on the Cholera Epidemic of 1866 in England* PP 1867–8:XXXVII:163; Charles Creighton, *History of Epidemics in Britain* vol 2 (second edition, 1965) 858.

**Table 4.2** Water quality and cholera, 1832–66. Source: *Rivers Pollution Commissioners Sixth Report: Domestic Water Supply of Great Britain* PP 1874:xxxiii:472.

| Year | Character of water | Cholera mortality per 10 000 population in London |
|------|--------------------|---------------------------------------------------|
| 1832 | Polluted           | 31.4                                              |
| 1849 | Very much polluted | 61.8                                              |
| 1854 | Less polluted      | 42.9                                              |
| 1866 | Much less polluted | 18.4                                              |

The cholera vibrio remains active for about 14 days in river water and is decisively affected by changes in temperature.[6] But, during its 'live' period, it is exceptionally dangerous: and this high degree of microbial activity is likely to be reflected in sudden and violent upswings in incidence and mortality as well as in unpredictable explosions following apparently decisive downswings. It is for these reasons that social historians, like a number of pioneering nineteenth-century epidemiologists, have analysed the time structures of cholera epidemics in order to ascertain the extent to which water transmission may or may not have been the dominant mode of spread.[7] Figure 4.4 shows the temporal pattern of mortality in London during the summer of 1849. The crucial turning point occurred during the third week of August. Throughout July weekly death rates had risen inexorably—from 152 to 339, from 339 to 678, from 678 to 783—to reach a peak of just under a thousand during the first week of August. At this point there was a brief respite, but between 12 August and 9 September the death rate leapt convulsively upwards again—a strong indication that the epidemic was now being more rapidly disseminated via public water distribution systems—and only finally restabilised at much lower levels during the last week of September.

The period between the first and second mid-century epidemics witnessed important changes in patterns of public consumption. More specifically, the decision by each of the water companies, with the exception of the Southwark, to improve the quality of its river sources, was to have a decisive effect upon the geographical incidence of the disease in 1853–4. But quantity as well as quality was involved: and as table 4.3 demonstrates, by the early 1850s a substantially larger proportion of Londoners were deriving their domestic water from company rather than non-company sources.

A number of conclusions may be drawn from the material presented in table 4.3. Depending upon income and geographical location,

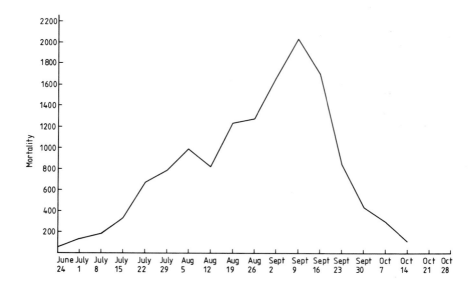

**Figure 4.4** Cholera mortality in London. June–October 1849. Source: *Report on the Cholera Epidemic of 1866 in England* PP 1867–8:XXXVII:163.

**Table 4.3** Water supply of London in 1849 and 1856 (average daily quantity in gallons). Sources: *Report of the General Board of Health on the Supply of Water to the Metropolis* PP 1850:XXII:6 and *Report of Select Committee on East London Water Bills* PP 1867–8:XXXVII:Appendix I: 374.

|                | 1849       | 1856       |
|----------------|-----------:|-----------:|
| New River      | 14 576 783 | 23 396 400 |
| East London    | 8 829 462  | 14 562 684 |
| Chelsea        | 3 940 730  | 5 532 000  |
| West Middlesex | 3 334 054  | 6 005 293  |
| Grand Junction | 3 532 013  | 5 478 361  |
| Southwark      | 6 013 716  | 10 331 122 |
| Lambeth        | 3 077 260  | 5 391 000  |
| Kent           | 1 079 111  | 2 680 000  |
| Total          | 44 383 129 | 73 376 860 |

Londoners at this time had access to four categories of drinking water: safe well water, unsafe well water, safe company water and unsafe company water. Had it been the case that all those who had drawn their water from wells in, say, 1850 had had access to safe company supplies by 1854, the incidence of cholera would certainly have been greatly reduced. But a substitution of this kind did not occur, and events in south London during the summer of 1854 demonstrated that large numbers of consumers had merely exchanged safe or unsafe well water for unsafe company supplies. The effect of the movement away from well water, together with intercompany differentials in terms of sources and purification techniques, was reflected in the geographical distribution of cholera mortality in south London in 1854—and this point is illustrated in the second large column of figure 4.5. The focus of infection within the capital as a whole was now indisputably located in the southern districts and, more specifically, among consumers of 'Southwark' rather than 'Lambeth' water. That water transmission did indeed play a crucial role in the dissemination of the epidemic of 1854 is shown in the time structure in figure 4.6. The general shape here is similar to that of 1849, with the death rate rising steeply week by

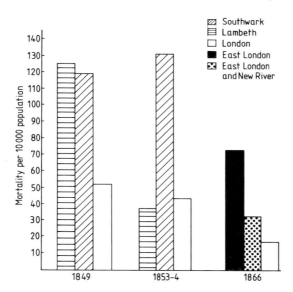

**Figure 4.5** Cholera mortality and water consumption per 10 000 population in London. Sources: *Twelfth Annual Report of the Medical Officer of the Privy Council* PP 1870:XXXVIII:616 and *Report on the Cholera Epidemic of 1866 in England* PP 1867–8:XXXVII:185.

week from 22 July to 12 August, slackening between 12 August and 26 August, but then spiralling upwards again to a higher peak in the week ending 2 September than at any comparable phase of the earlier epidemic. Thereafter, there was an unusually slow downswing with mortality only declining to decisively lower levels in the week beginning 7 October.

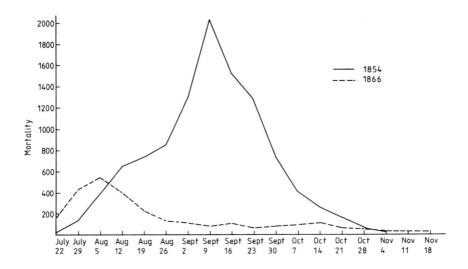

**Figure 4.6** Cholera mortality in London, June–November 1854 and July–November 1866. Sources: *Report on the Cholera Epidemic of 1866 in England* PP 1867–8:XXXVII:150 and *Ninth Annual Report of the Medical Officer of the Privy Council* PP 1867:XXXVII:336. Note that the dating for the two years was not identical—in this figure 21 July 1866 is equated with 22 July 1854.

The role which the Southwark Company played in the spread of the disease is fully revealed in figure 4.5, which is based on the pioneering fieldwork of John Snow and John Simon. The initial investigation was carried out by Snow in collaboration with the Registrar-General's office, and his conclusions were subsequently refined by John Simon. Simon examined 31 sub-districts—8 of them 'monopolised' by a single company, and the other 23 'shared'. For 1854, he was able to obtain accurate figures for the size of population in the whole region, as well as for people and houses in the supply area of each of the companies. But, for his calculations for 1849, Simon was forced to make the assumption that the number of houses supplied by the two companies was one eleventh lower than in 1854. The essentials of this exercise are summarised in tables 4.4 and 4.5 which show mortality in 45 streets, containing 3034 houses, which were 'divided' absolutely equally between the two companies, and comparative mortality within the total populations supplied by the two companies in 1849 and 1854.

**Table 4.4** Mortality from cholera per 10 000 population in 45 south London streets, 1849 and 1854. Sources for tables 4.4 and 4.5: *Report on Last Two Cholera Epidemics as affected by Consumption of Impure Water* PP 1856:LII:384–5 and *Twelfth Annual Report of the Medical Officer of the Privy Council* PP 1870:XXXVIII:613–17.

| Origin of water supply | 1848–9 | 1853–4 |
|---|---|---|
| Lambeth | 164 | 57 |
| Southwark | 95 | 164 |
| Unknown | 26 | 21 |

**Table 4.5** Mortality in areas supplied by south London water companies.

| Origin of supply | Population | No of houses | Deaths | Rate per 10 000 |
|---|---|---|---|---|
| Lambeth | | | | |
| 1848–9 | 151 732 | 22 594 | 1925 | 127 |
| 1853–4 | 166 906 | 24 854 | 611 | 37 |
| Southwark | | | | |
| 1848–9 | 243 791 | 21 776 | 2880 | 118 |
| 1853–4 | 268 171 | 23 976 | 3476 | 130 |

In 1866 mortality from cholera in London accounted for a considerably lower proportion of the national aggregate than it had in 1854— 5548 out of 14 378 as compared with 10 684 out of 20 097. The disease during this final outbreak was very heavily concentrated in the East End, a point which is endorsed in the third large column of figure 4.5. Hackney, St Giles, Shoreditch, Lambeth and Greenwich all recorded above average death rates compared with London as a whole, but by far the largest numbers of fatalities occurred in Bethnal Green, Whitechapel, St George-in-the-East, Stepney, Mile End and Poplar. Yet even in these devastated districts mortality was lower than it had been in south London in 1849 and 1854. (The death rate per 10 000 population in the most seriously affected district in 1866—Stepney—was just below 120; in 1854 in Bermondsey it had been in the region of 180; and in Rotherhithe in 1849 it had risen to the horrific figure of 200.) The third large column of figure 4.5 shows that, although water delivered by the East London Company was clearly the decisive trigger for the epidemic as a whole, New River supplies, drawn from similar sources in the River Lea, were also probably contaminated. Neither source, though, was as grossly polluted for so extended a period as those of the major south London concerns in 1849 and 1854. The time structure of the epidemic of 1866 is shown in figure 4.6. Differences both in scale, shape and fluctuation between this and the epidemics of 1849 and 1854 may be observed but the characteristic explosion—although it occurred earlier in the summer of 1866 than in either 1849 or 1854—was common to all three outbreaks. The steep upward gradient between late July and the end of August is complemented by the downward movements thereafter: but the wave-like tremors at relatively low levels of mortality between the beginning of September and the middle of October suggest minor reactivations of water-carried infection. Clearly, good weather in late summer and early autumn, as well as the deadly incompetence of the water companies, could kill the poorest of London's inhabitants.

We may now trace the way in which river and water pollution gradually came to be accepted as primary determinants of cholera. In essence, this was a process which involved the erosion of the body of thought which constituted the most powerful epidemiological ideology of the early and mid-nineteenth century, the miasmatic theory of disease.[8] The dominant explanation which had come to prevail following the epidemic of 1832 was that cholera had its origins in a nexus of adverse atmospheric conditions, inadequate and unsuitable food, and an intemperate and ill-disciplined way of life. Immediately before the second outbreak in 1848–9, the Metropolitan Sanitary Commission, directing its advice predominantly at the poor, warned that 'where people live irregularly, or on unsuitable diet and at the same time filthily' they must expect to perish.[9] If, however, those subjected to noxious vapours behaved in a moral and abstemious manner and paid more attention, in particular, to cleanliness, they might well survive an epidemic onslaught. Yet, in the aftermath of the slaughter of 1849 it was difficult, indeed impossible, to point to poor districts which had in fact avoided decimation as a result of any heightened collective awareness of hygiene. And it was the absence of proof of this type, together with the inherent ideological loading of advice offered by middle-class reformers to the working class, which dictated that a profound moralism became embedded within the miasmatic model. According to this orthodoxy, personal moral strength and resilience in adversity, rather than external environmental processes, dictated who would and who would not survive an epidemic. Indeed, the very term 'pollution' carried, as it still does carry, a set of meanings which were subjective and normative rather than objectively socio-environmental.[10] Thus a representative of the governors of the House of Correction at Coldbathfields defined what seemed to him to be the predisposing factors of cholera as 'the absence of cleanliness, of decency and all decorum—the disregard of any heedful separation of the sexes—polluting language and scenes of profligacy . . .'.[11] This was the 'immoral' environment in which, given adverse miasmatic conditions, epidemic disease was bound to thrive.

It had, however, long been recognised, as basic topographical fact, that cholera attacked riparian communities with exceptional ferocity.[12] And, in its analysis of the epidemic of 1849, the Board of Health dealt with this phenomenon in a manner which directly confronted the issue of river pollution. 'It is difficult to arrive at any other conclusion', the Board stated, 'than that streams polluted by the refuse of large masses of people so deteriorate the air as to operate in the time of a destructive epidemic, when all depressing agents have increased force, injuriously on the human frame, and thereby predispose it to the attacks of disease.'[13] According to this variant of miasmatic theory, a great river such as the Thames might become in its grossly polluted state a purveyor of atmospheric infection. This was an interpretation which meshed with William Farr's growingly influential theory that there was a connection between height above sea-level and relative immunity from cholera: a suburb like Hampstead, so this argument ran, was only minimally affected by the disease because it was geographically distanced from and meteorologically resistant to the river-borne vapours which hovered over the inner-city areas.[14] The members of the two committees responsible for investigations into the outbreak of 1853–4 were conversant with John Snow's preliminary field studies and were aware of the interest which they had provoked. The role which unsafe water may or may not have played was therefore more exhaustively examined than it had been in the aftermath of the epidemic of 1848–9. And yet what is most striking about the account prepared on behalf of 'professional' medicine, *The Report of the Medical Council*, is that it yet further expanded the parameters of miasmatic doctrine in order to accommodate both the observed data and the epidemiological logic and structure of Snow's innovations. River pollution, according to this view, could indeed exacerbate an 'epidemic atmosphere': and there was no doubt that 'the foul state of the Thames and its share in rendering the atmosphere impure . . .' had played a part in precipitating the recent attack. But medical orthodoxy demanded that this observation be closely linked to the type of meteorological analysis which had first been popularised in 1849. 'A great excess in the temperature of the Thames at night, as compared with that of the superincumbent atmosphere' was therefore interpreted as 'giving rise to nocturnal clouds of vapour, which are necessarily charged with impurities derived from the foul contents of the river. . . .'[15] As for the possibility of cholera being directly transmitted via polluted water supplied during the notorious Soho tragedy or in the districts to the south of the river the response was negative: the phenomena documented by Snow were simply reintegrated into a predominantly miasmatic schema. 'The suddenness of the outbreak', the committee concluded, 'its immediate climax, the short duration, all point to some atmospheric or other widely diffused agent still to be discovered and forbid the assumption, in this instance of any communication of the disease from person to person, either by infection or by contamination with the excretions of the sick.'[16] Events which, in John Snow's view, seemed strongly to support the hypothesis that water had been primarily involved in the spread of the epidemic, were here deployed as yet further confirmation of the underlying primacy of atmospheric factors. Unsafe water may well have aggravated the severity of the attack 'but it had not acted as a transmitting medium'.[17]

It might be thought that the other body which examined the epidemic of 1853–4, the Committee for Scientific Inquiries, which counted John Snow and William Farr among its members, would have been more aware of the epidemiological implication of river pollution. But this was not the case. The fundamental process was argued to be 'organic decay', occurring either in air or water, inside or outside the body.[18] Given this framework, the committee held that it would have been premature to endorse what was becoming increasingly widely known as the 'exclusive waterborne theory': and in order to substantiate the continuing centrality of atmospheric processes, an appeal was again made to allegedly irrefutable meteorological evidence. 'For on the supposition', it was claimed, 'that the choleraic infection multiplies rather in air than in water, meteorology explains how the balance of healthfulness is weighted in favour of the higher levels, by their less participation in the high night-temperature of the metropolis, by their comparative clearness from mist, and above all by the curative resources of more free ventilation.'[19] Thus it was that an ever more flexible version of the atmospheric theory was reaffirmed, with only the eclectic William Farr moving significantly closer to Snow's position. 'The cholera matter or cholerine', Farr now believed, 'where it is most fatal is largely diffused through water, as well as through other channels.'[20]

The years between the third and fourth cholera epidemics were characterised by important changes in patterns of metropolitan water supply. The extent to which these changes were precipitated by theoretical innovation, and more particularly by John Snow's analysis of the spread of the disease in the capital in 1849 and 1854, was minimal. But Snow's work is nevertheless central to an understanding of water and river pollution as discrete environmental and epidemiological problems in the mid-nineteenth century and to the way in which these problems were perceived by those professionally concerned with public health and the prevention of disease. It has already been seen that earlier investigators had noted that cholera seemed to travel with unusual alacrity along river courses, and that observations of this type were later accommodated within the miasmatic paradigm. John Snow's approach was quite different. 'Rivers', he wrote in 1854, 'always receive the refuse of those living on the banks, and they nearly always supply, at the same time, drinking water of the community so situated . . . the water serves as a medium to propagate the disease amongst those living at each spot and thus prevents it from dying out through not reaching fresh victims.'[21] It had also been believed that, although unsafe water might 'predispose' an individual to disease, it was incapable of transmitting cholera from person to person. Snow now subjected the concept of 'predisposition' to rigorous scrutiny. 'Many medical men', he wrote, 'whilst they admit the influence of polluted water on the prevalence of cholera, believe that it acts by predisposing or preparing the system to be acted upon by some unknown cause of the disease existing in the atmosphere or elsewhere . . . if the effect of contaminated water be admitted, it must lead to the conclusion that it acts by containing the true and specific cause of the malady.'[22] Three important components of the miasmatic credo—the rationale for the rapid movement of cholera along water courses, the place of unsafe water in the hierarchy of 'predisposing' causes, and the concept of 'predisposition' itself—were now under threat. And cholera was now also being defined as a species of dynamic living matter, in constant interaction with the environment, including man. 'I consider', Snow insisted in a classic statement before a select committee in 1854, 'that the cause of cholera is always cholera; that each case always depends upon a previous one.'[23]

The idea, moreover, that diseases were interchangeable and that, depending on subjectively-willed resilience and moral self-control, an individual might either vanquish or be destroyed by perversely unpredictable forms of 'fever', was now in jeopardy. The specificity of disease could not yet be demonstrated: but, by concentrating on the inconsistencies of the existing orthodoxy, Snow had simultaneously indentified and objectified the problem of unsafe water. And in the longer term, this would both lighten the heavy psychological burden which miasmatic doctrine decreed that those subjected to the threat of epidemic disease must labour under, and bring river pollution, and its prevention, more fully into the domain of controlled inquiry.

By 1866 a small minority of specialists in public health, notably William Farr and Edward Frankland, had become convinced that an outbreak of cholera should be taken as an indication that water supplies were highly suspect and river sources indefensibly polluted. Consensus professional opinion, though, continued to be antipathetic to the major tenets of the 'exclusive' water-borne explanation of the spread of epidemics and of the 20 or so metropolitan medical officers of health whose views have been sampled, only two can be said to have supported a position which gave high priority to unsafe water. Two years before the epidemic of 1866, William Budd, who had revolutionised the study of typhoid, summarised professional attitudes towards infectious disease in the following terms: 'A very large, and by far the most influential school in this country, a school which probably embraces the great majority of medical practitioners and the whole of the 'sanitary' public holds . . . and teaches that sundry poisons are constantly being generated *de novo* by the material conditions which surround us.'[24] A majority of medical officers concurred with this account of things in the mid-1860s and continued to relegate unsafe water to a subsidiary position. Few were as hostile as the acting officer for Mile End Old Town who claimed in 1866 that he had 'never seen or read a single reliable fact' to support the water theory[25] but a number of his colleagues were ready to deploy explanations which implicitly minimised the direct influence on the incidence of cholera of unsafe water derived from polluted rivers.

The most frequently cited proposition, and one which dissenting contemporaries found exceptionally difficult to refute, was that numerous 'control' groups in institutions such as schools, asylums and workhouses had been observed to drink allegedly infected water during the epidemic of 1866 but had remained immune from cholera.[26] Less logically secure, but of continuingly wide appeal, were empirical observations which emphasised the counterbalancing effects of poverty, class differentials and a dissolute subjective life style—all of them factors thought to outweigh the effects of unsafe water.[27] But by far the most popular total explanation centred on the well-tried hierarchy of 'bad water, bad air, defective drainage, overcrowding, dirty and irregular habits'[28], with the important rider that it was invariably the traditional 'miasmatic' conditions and not unsafe water *per se* which occupied the determining position. By the mid-1860s elements of Pettenkofer's 'soil' theory were also being built into the apparently indefinitely expandable parameters of the miasmatic paradigm. 'There certainly is a difference between the mortality of the parts supplied by the three companies', reported the medical officer for Kensington, tending towards this interpretation, 'but I do not attribute this at all to the water, but rather to the drainage, crowding and more especially the clay soil of the northern part.'[29] Nor was there any diminished commitment to the 'Thames-borne' theory of infection which, following William Farr's work on elevation, meteorology and the 'epidemic atmosphere', had gained considerable support in 1849 and 1854. 'The present epidemic', explained the medical officer for Greenwich, who was clearly attracted to this type of theoretical framework, 'has mainly existed in all such parts of my district as are contiguous to the river Thames. The nearer the river the more cases of cholera, and the greater the severity of attacks generally, the disease gradually decreasing in virulence and numbers as the distance increases from the river.'[30] It was Thomas Orton, a combative opponent of the 'exclusive' water theory, who provided the most comprehensive guide to the miscellany of explanations which claimed substantial numbers of supporters in the public health movement in London in 1866. 'Pretty generally amongst all classes the theory of the water poison is repudiated', Orton wrote, 'especially among the poor, who have chiefly felt the shock: and this opinion is commonly shared by professional men.'[31] He went on to compile a catalogue of explanations which was more variegated than even the rash of theories which had been put forward in 1849 and 1854. Distinct and separate atmospheric, telluric, electric and ozonic influences comprised the first group of causal accounts. There followed a grouping, illustrating the impact of the work of Liebig and Pasteur, and emphasising the primacy of 'zymotic fermentation', 'spread by human communication, animalicular bodies, specific poisons and fungi'. Finally, at the bottom of the list, and deliberately separated from other 'poison'-based theories, Orton noted the modes of infection associated with John Snow and William Budd.[32]

It might be concluded from this summary description that the metropolitan medical officers of the mid-1860s had no interest in water quality: but this was not the case. Given the existing consensus on the transmission of infectious disease, it was predictable that it would be the issue of quantity rather than quality which claimed the greater attention—if houses and bodies could be kept clean, then the general death rate would decline. But the companies' refusal to extend constant supply and their Draconian policies towards poor tenants who either misunderstood or failed to comply with a by now complex body of regulations—or who fell behind with their payments—held back progress. At this point, as we have already seen, the issues of quantity and quality in the transmission of water-borne disease became intertwined and inseparable. For, where dwellings or even whole courts were dependent on either a common butt or an inadequately serviced cistern, 'domestic pollution' would be rampant. One of the attractions to the medical officers of this explanation of the process of infection was that it could be accommodated both within the miasmatic and an inexplicitly formulated version of the 'water theory'. According to the former, emanations from an unsafe cistern interacted with atmospheric impurities to produce disease, while according to the latter, cholera or typhoid were quite simply 'swallowed'. Thus both before, during and after the epidemic of 1866 a majority of medical officers campaigned for constant supply, and pending such an extension, for compulsory inspection and regular cleansing of domestic storage systems.[33] And such a campaign could be conducted without violation of a total belief system which still resisted full incorporation of the 'exclusive' water theory.[34]

During the epidemic itself, the medical officers kept in contact with one another through their professional association and via the Registrar-General's office. But communications of this type remained sporadic and uncoordinated and were of limited importance when compared with the close relationship which grew up between William Farr at the Registrar-General's office and Edward Frankland in his chemical laboratory. Previous investigations by Snow, Budd and Farr had used *post hoc* statistical analyses to evaluate the importance of polluted water in the spread of cholera. By the mid-1860s, however, knowledge among the medical avant-garde about the connections between river pollution, water supply, and patterns of mortality according to registration district and what Farr termed the 'water fields' of the companies allowed more rapid—although still not nearly rapid enough—environmental diagnosis.

By the end of July 1866, with mortality in the East End reaching traumatic levels, Farr was well aware of the possible role of river pollution in the transmission of the epidemic. At this juncture, however, he was pessimistic about the possibilities of preventive action. 'The Company will, no doubt, take exemplary pains to filter its water', he confided to Frankland on 28 July, 'but it is not easy to guarantee the purity of water drawn from such a river as the Lea, in dangerous proximity to sewers, cuts and canals.'[35] On 30 July, in another letter to Frankland, he reiterated his contention that there must be a connection between the abnormal contamination of the Lea and the spiralling mortality from cholera in the East End, and by 31 July he was noting that the epidemic 'quite reminds me of the Southwark slaughter'.[36] Rapid action was imperative but, when approached informally, the East London Company bridled at the suggestion that it might be even minimally responsible for dissemination of the epidemic. On 2 August, in an attempt to stifle damaging rumours, Charles Greaves, the company engineer, who was to play a dramatic role in the subsequent unravelling of the tragedy, wrote to *The Times* giving an assurance that the water which was being drawn from the Lea was absolutely safe.[37] This was in partial response to Farr's statement that use was still being made of a canal which connected the company's filter beds at the Old Ford works directly to the river. Unknown to Greaves, Farr had already examined detailed maps and, to his intense surprise, had discovered that there were in fact two pumping establishments—one at Old Ford and the other at Lea Bridge. At the former, besides a covered reservoir, there were also two others which were not covered. Whatever the means of transmission of the East End cholera, then, the company had clearly contravened a clause of the Act of 1852 which outlawed the use of uncovered reservoirs within five miles of St Paul's.

Greaves' initial response had been forthright. He accused the Registrar's office of consulting out of date maps and went on to insist that, although the canal in question had yet to be filled in, 'not a drop of unfiltered water has for several years been supplied by the company for any purpose'.[38] But the existence of the uncovered reservoirs had already partially undermined his position and now Farr moved swiftly. On 3 August he wrote to Frankland's assistant, 'The engineer, Mr. Greaves, states that there is still a connection between the *wells* of the engine at *Old Ford* and the uncovered reservoirs, but denies that these waters are ever used.'[39] To test the reliability of this assertion, the assistant was requested to undertake analyses of water from the single covered reservoir, as well as from the two uncovered reservoirs at Old Ford, and then at the Lea Bridge works both before and after filtration. These chemical tests revealed no significant differentials—and the repercussions which were to flow from this unsuccessful effort to identify the presence of the water 'poison' were to prove exceptionally significant in the ensuing debate.

Meanwhile, on 4 August, with the death toll still horrifically high, Frankland suggested that, as an emergency measure, all suspect water supplied by the East London Company should be treated with permanganate of potash before being passed through a double rather than a single filter. The extra bed was to consist of the kind of charcoal which was usually to be found in domestic cisterns. But Frankland's status was still ambiguous—that of an informal though respected 'outsider'—and, had they yielded to his request, company officials might have involved themselves in a tacit admission that their water was in fact fundamentally implicated. Greaves, therefore, rejected Frankland's advice on the grounds that the double system would prove unduly costly.[40]

In terms of argument and counter-argument on the role of water quality the company held its own, both during and immediately after the virulent but brief epidemic. Authoritative, alternative miasmatic

and 'sociological' explanations, accounting for the particular vulner-
ability of the East End, were plentiful. Thus N Beardmore, engineer to
the River Lea Trust, wrote to the Registrar's office on 6 August to insist
that 'overcrowding, deficiency of drainage, and inferior articles of food
are more likely to have promoted cholera than impurity or deficiency of
water'. The East End was, moreover, populated by 'dock labourers,
sailors, mechanics in the new factories, and great numbers of
laundresses'—all social groups, so Beardmore argued, whose underdeveloped sense of hygiene made them especially susceptible to
disease.[41] An even stronger defence was that, according to the best of
available chemical methods of analysis, the East London Company's
water had been found over the previous few years to compare favour-
ably with that of the Thames companies.[42] And, as the epidemic waned,
Edward Frankland himself admitted the extreme technical difficulty of
making an unequivocal association between water pollution and cholera
incidence.[43]

Throughout the numerous subsequent investigations into the
epidemic the East London Company defended itself sturdily and
unapologetically. The company was especially fortunate in being able
to call upon the services of Henry Letheby, one of the most able and
logical critics of the 'exclusive' water-borne theory. Letheby pointed to
the technical and methodological difficulties of locating the cholera
'poison'.[44] He also made telling use of the by now familiar argument
that 'teetotallers and others who drank largely of the East London water
in its unboiled condition' had been 'signally exempt from the disease'[45],
and that it would have been equally convincing (or unconvincing) to
forward the view that the epidemic had been transmitted by a common
gas supply.[46] He also dwelt at length on the implications of 'closed'
institutions where there had been heavy consumption of East London
water but relatively low mortality, and defied supporters of the
'exclusive' theory to explain this apparent contradiction.[47] But the
company's case was fatally weakened by the sudden and apparently
unabashed volte-face in December 1866, on the part of the engineer,
Charles Greaves, who admitted that, especially during periods of high
demand, there had been an 'implied sanction' to draw upon unfiltered
water. 'They [the canal works]', he confessed to the Rivers Pollution
Commissioners, 'were abandoned because the system of filtration, as a
matter of course, put them out of use, then they were not applicable to
any other purpose and I thought myself more justified in keeping them
[the uncovered reservoirs] full than I was in emptying them, considering
that the question of quantity was at times of such importance with
reference to accidents, that it was advisable to keep reservoirs which
were already constructed and which I could not apply to any other
purpose, to meet such necessities as might arise.'[48] But in exoneration—

and this was to prove significant—Greaves insisted that Letheby had assured him that the water stored for delivery was of a high standard; and Frankland, who had performed a nearly contemporaneous analysis, had been unable to contradict this assessment.[49]

The Board of Trade report was harsh but, from the company's point of view, not wholly without redemption. The company was found guilty of distributing unfiltered water—'a distinct infringement of the provisions of the 1852 Act'.[50] Water had been allowed to flow directly from the uncovered to the covered reservoir; there had been confirmed seepage of unsafe supplies into previously filtered water; and raw Lea water had also found its way into supplies prior to distribution. But the Board of Trade inspector, Captain Tyler, was not prepared to enter a verdict of unqualified guilt. The crucial rider was phrased in the following terms: 'any poison so distributed would have been in a condition, if it were soluble in water, of considerable dilution; and I am not prepared on that account, as well as in consideration in other respects of their district, to go as far as the Memorialists, in asserting that this water was the "principal" if not the sole cause of the fearful mortality from cholera'.[51] In other reports concerned with the distribution of blame for the outbreak the 'deplorable state' of the East End would be made to carry a much heavier burden. Thus the Rivers Pollution Commissioners, while deploring the company's laxness in distributing filtered supplies which had then immediately been subjected to seepage by untreated water from the Lea, drew particular attention to 'local conditions'. The 'low level of the district, the use of polluted wells, the saturation with sewage of the subsoil, and the excessive accumulation of stagnant sewage in ditches and cuts arising in great measure from the storm over-flows of the metropolitan sewage works' were presented as the primary conditions in which the disease had prospered.[52] No objection could be made to locating the epidemic in its full environmental context; but there was an implicit view of the causation of cholera at work here which could easily be modified in such a way as to very nearly wholly exonerate the company.

The initial report of the Select Committee on East London Water Bills, which had been drafted in the immediate aftermath of the epidemic, had noted the committee's 'entire concurrence' with the Board of Trade's castigation of the company under the legislation of 1852.[53] Yet the committee had subsequently allowed itself to be impressed by the evidence of a number of witnesses who were anxious both to minimise the company's direct responsibility for the calamity and to convince the public that further capital expenditure was essential if increased demand were to be satisfactorily met: investors' confidence had to be restored. So, by ways and means that are not clear, the draft report was revised and published in a form which substantially demoted the role of unsafe water. 'We think it right to observe', the committee concluded, 'that the evidence leads to the opinion that the spread of cholera might equally be ascribed to defective sanitary arrangements and to other causes.'[54] The East London Company now counter-attacked with a degree of confidence. In July 1867 the directors wrote to the Board of Trade seeking an official pardon and disputing the implications which had been drawn from the inspector's report by the public. Carefully preselected phrases which Tyler, for reasons of accuracy, had allowed to stand in a deliberately tentative form, were lifted from their context and skilfully used to give the impression that the company had been misrepresented and unjustly accused.[55] The directors minimised the misdemeanours of 1866, while simultaneously congratulating themselves on undertaking improvements which were, they claimed, neither legally binding nor necessary in terms of the maintenance of public health. They also pointed out that the three major shortcomings—the direct flow of water from the uncovered to the covered reservoirs, the seepage of unsafe water into previously filtered supplies, and the direct infiltration of water from the Lea—had been rectified. 'Observance of all these remedial proceedings', they protested, 'has been secured by heavy penalties voluntarily imposed, and the Company will of course perform what it has in this respect undertaken; thus incurring a very considerable outlay, not perhaps necessary for the remedy or prevention of any actual mischief, but, at all events, desirable as cautionary measures, and for quieting the public mind.'[56] But, following a forth-right memorandum from Captain Tyler, the department declined to reconsider the decision that there had been a serious infringement of the Act of 1852.[57]

By late 1867 public and specialist opinion had hardened against the company. William Farr had spoken with deliberate ambiguity of 'this disastrous accident in East London'.[58] Now the invariably moderate *British Medical Journal*, which continued to support a broadly miasmatic theory of cholera transmission, was incensed when it unearthed evidence which showed that on several occasions in 1864 and 1865, as well as in 1866, the company had delivered unfiltered water.[59] 'It is true that supervision by the State lessens private responsibility', commented the journal's editorial, but 'is the tremendous responsibility which rests upon them sufficiently brought home to our water companies? Who has been brought to account for the terrible result of the day's work at East London?'[60] According to the more militant *Lancet* the company had been shown to be guilty of taking advantage of the scientific and technical ambiguities which still bedevilled efforts to relate water pollution to outbreaks of cholera. 'One cannot be otherwise than pain-fully impressed . . . with the disadvantage at which the public are placed in regard to a matter so vitally affecting them as the purity of their water supply, when the Companies, in whose hands that supply is a monopoly, secretly infringe the law, trusting to the difficulties by which discovery is virtually rendered next to impossible.' It was 'greatly to be regretted that a heavy penalty has not been levied for the infraction of the law'.[61] Among specialists, John Simon urged that the companies ought finally to be subjected to more intensive and more fully informed public surveillance. 'The power of life and death', he wrote, 'in commercial hands is something for which, till recently, there has been no precedent in the world, and even yet the public seems but slightly awake to its importance.' But, significantly, Simon, like many other public health specialists, went on to deplore the failure of existing techniques of chemical analysis to provide authoritative correlations between an epidemic and the cholera 'poison' in a given source.[62] It was for this reason that when he directly addressed the scientific community, Simon was cautious in his assessment of the cause of the epidemic. The most outspoken critique of the behaviour of the East London Company had been prepared, for Simon's department at the Privy Council, by Netten Radcliffe and, in a characteristically cautious addendum to Radcliffe's account, Simon asked rhetorically, 'For the substance of Mr. Radcliffe's conclusion, is it necessary to assume that water was drunk?' He then added 'if the particular form of the conclusion be set aside, and the conclusion simply affirm that the distribution of the Old Ford water caused the east London outbreak of cholera, then, to whatever extent that distribution operated on and through the soil of the territory, so far Pettenkofer's doctrine may admit of more or less plausible application to explain the partialities of the attack'.[63]

Simon, scrupulous scientist that he was, would not accept hypotheses which had not been verified either under laboratory conditions or through irrefutable statistical and epidemiological research: his strength lay precisely in his ability to reconcile 'progressive' and 'consensus' opinions in medicine and public health practice.[64] Thus, while he could appreciate the strength of the 'exclusive' water theory, the balance of empirical evidence still seemed to him to favour an overall interpretation cast in terms which could be accommodated within the 'soil' theory of infection. Scientifically, such a position was unobjectionable: it was always likely, however, to be exploited by those who were less disinterested than Simon or who might be tempted to oversimplify the soil theory itself.

Netten Radcliffe, on the other hand, was aware of the need to refute prevailing doctrine in order to dramatise the dangers of river and water pollution and the laxness of the water companies. The two most convincing counter arguments to the 'exclusive' theory, which had been coherently stated by Henry Letheby, were, first, that there was no known chemical means of locating the cholera 'poison'; and, second, that large numbers of consumers of allegedly unsafe supplies had nevertheless escaped infection. Radcliffe's rebuttal of this widespread and influential mode of argument was of great importance in the development of public health surveillance. 'It is argued', he reported, 'in respect of a serious objection to this [the 'exclusive'] theory, arising out of the actual or relative immunity from cholera of certain districts and institutions supplied with the cholera-infected water, that in the present state of our knowledge of the outbreak, the positive and generally more applicable facts may justly, and for practical purposes, warrant a conclusion apparently in contradiction with certain negative facts of much more restricted application.'[65] If epidemic water-borne diseases were to be eradicated, it was a necessary and binding, and not simply a highly desirable, condition that water drawn from unsafe river sources should be subjected to systematic and controlled treatment. 'For not only is it now certain that the faulty water supply of a town may be the essential cause of the most terrible epidemic outbreaks of cholera, typhoid fever, dysentery and other allied disorders: but even doubts are widely entertained whether these diseases, or some of them, can possibly attain general prevalence in a town except where the faulty water-supply develops them.'[66] That Radcliffe was expressing a minority view even within the atypical consensus shared by the vanguard of the public health movement requires no further emphasis. In the absence of tests for specificity, of a generally agreed *modus operandi* for the inferred cholera 'poison', as well as knowledge of the precise bacteriological effects of filtration on raw river water, the apparently deceitful obduracy of the East London Company can be more easily comprehended. Nor need this evaluation be interpreted as exoneration, for the directors and the engineer did not hesitate to exploit existing medical ideologies both to flout the law and, later, to justify actions which the law condemned.

It is, however, necessary to be clear about both the nature and the aims of the legislation which had been passed in 1852 as well as of general attitudes towards water-transmitted disease in the mid-1860s. It is tempting to assume that the clauses which condemned the use of water from uncovered reservoirs and which insisted upon 'adequate' filtration and storage were devised specifically to prevent epidemics of what we now know to be exceptionally dangerous water-transmitted infections. But a major aim of this chapter has been to insist that only a minority of medical experts, and scarcely anyone at all in a position of authority within the London water companies, really believed that polluted supplies could precipitate deadly epidemics of cholera and other infections. Pure water was desirable and it could certainly help to protect the body against serious illness but it was emphatically not conceived as an essential—if not the essential—precondition for exemption from cholera. Broad associations between unsafe and/or unfiltered water and generalised ill-health were widely accepted but causal connections were only rarely made and were in any case more likely to be perceived intuitively by ordinary people than propounded by experts. In that respect, the east Londoner who found a dead and decomposing eel in his supply pipe in 1866 and moved immediately to the conclusion that the East London Company had been responsible for the dissemination of cholera was closer to a correct analysis than the miasmatists or those who believed in the 'soil' theory of disease.[67]

In terms of the ideas which it propounded on the connection between river pollution, unsafe company water and the transmission of cholera, William Farr's classic report on the epidemic of 1866 was more extensive and less circumspect than the official investigations of the outbreaks of 1849 and 1854. Influenced both by the 'germ' theories associated with Budd and Snow and the chemical hypotheses of Edward Frankland, Farr believed that the East End had been 'poisoned' by its major water company and that the state, or some other agency, must now take steps to ensure that such a tragedy would never be repeated. River pollution and its appalling epidemic potential could no longer be accepted as 'facts of nature': social and political remedies must be immediately and imperatively sought. As always in Farr's work, there was much that was tendentious—notably page upon page of bizarre statistical rumination[68]—but this was outweighed by an insistent desire to ascribe responsibility where responsibility was due: and to point to the feebleness of existing legislation which gave Londoners no control over a fundamental commodity, that of safe drinking water.

Edward Frankland's attitude towards the events of 1866 was more complex. In terms of the prevention of disease and the measures which must be taken to ensure that so monstrous an event would never recur he was in agreement with Farr. But, as a scientist, he had become increasingly sceptical of the possibility of detecting the water 'poison' by chemical means: and, as an expert in public health, he could foresee comparable epidemics in the future when he and Farr—or their successors—would be attempting to locate a deficiency in purification procedures when the slaughter would already be mounting. Frankland was a pessimist: and the disaster of 1866, when he had known that something had gone horrifically awry, but had been unable to lay bare the underlying epidemiological processes, would stay with him for the rest of his life.

Although the demographic impact of cholera in nineteenth-century Britain was small, its social and medical significance can hardly be exaggerated. This was a disease which, because of its terrible and apparently uncontrollable symptoms, seemed both to reflect and threaten the fragile stability of early and mid-Victorian urban society. The manner in which it spread, and how it might be prevented, divided reformers and scientists, and reduced the public's belief in the effectiveness of medical intervention. But the quest for the epidemiological and social meaning of cholera led, albeit slowly, to the identification of unsafe water and river pollution, as major determinants of epidemic disease. And in that sense the Victorian obsession with the infection and its causation was central to the full emergence of pollution as a major social problem in the newly industrialising society; and integral, also, to that mode of thought which would, by the end of the century, come to be recognised as distinctively 'environmental'. It has been suggested that London constituted a vast social laboratory in which new ideas on cholera and the environment were forged and tested between the 1840s and the early 1870s. But what has also been strongly implied in this chapter is that, within that laboratory, there was no inevitable or incremental movement from 'ignorance' to 'enlightenment'. Indeed, as the events and attitudes of 1866 reveal with savage clarity, when public health specialists attempted to come to terms with the idea of river pollution and water-transmitted disease at this time they were often deeply confused: uneasily becalmed, as it were, between the entirely credible, generalised certainties of miasmatic orthodoxy and the crystalline structures of the germ theory of disease.

1    Norman Howard-Jones, 'Cholera Therapy in the Nineteenth Century', *J. Hist. Med.* **27** (1972) 373–95. For case fatality rates see G S Wilson and A A Miles (eds), *Topley and Wilson's Principles of Medicine and Immunity* (fifth edition, 1964) vol II, 1726.

2    Hugh Paul, *The Control of Diseases Social and Communicable* (second edition, Edinburgh and London, 1964) 345; and Wesley Spink. *Infectious Diseases: Prevention and Treatment in the Nineteenth and Twentieth Centuries* (Folkestone, 1978) 162.

3    For an assessment of such effects in the East End in 1866 see Gareth Stedman Jones, *Outcast London: A Study in the Relationship between Classes in Victorian Society* (Oxford, 1971) 102–3.

4    The medical profession was widely portrayed as a willing accomplice in such plans. See Asa Brigs, 'Cholera and Society in the Nineteenth Century', *Past and Present* **19** (1961) 88; and S E Finer *The Life and Times of Sir Edwin Chadwick* (1952) 349.

5    R Thorne Thorne, *The Progress of Preventive Medicine during the Victorian Era* (1888) 59.

6    Martin Alexander, *Microbial Ecology* (New York, 1971) 351.

7   See the analysis in M Durey, *The First Spasmodic Cholera Epidemic in York 1832* (York, 1974) and the comments in Lambert, *Sir John Simon 1816–1904 and English Social Administration* (1963) 128.

8   The miasmatic theory and the conflict between contagionism and anti-contagionism in the early nineteenth century have given rise to a wide-ranging and controversial literature. Mention may only be made here of Erwin H Ackerknecht, 'Anticontagionism between 1821 and 1867', *Bull. Hist. Med.* **22** (1948) 562–93; Margaret Pelling, *Cholera, Fever and English Medicine, 1825–1865* (Oxford, 1978); Roger Cooter, 'Anticontagionism and History's Medical Record' in P Wright and A Treacher (eds), *The Problem of Medical Knowledge* (Edinburgh, 1982) 87–108. These should be supplemented by national and regional studies of cholera and, especially, Charles Rosenberg, *The Cholera Years: The United States in 1832, 1849 and 1866* (Chicago, 1962); R J Morris, *Cholera 1832: The Social Response to an Epidemic* (1976); M Durey, *The Return of the Plague: British Society and the Cholera 1831–2* (Dublin, 1979); and A A MacLaren, 'Bourgeois Ideology and Victorian Philanthropy: the Contradictions of Cholera' in his edited collection of essays, *Social Class in Scotland: Past and Present* (Edinburgh, 1976) 36–54. The contextual importance of J M Eyler, *Victorian Social Medicine: The Ideas and Methods of William Farr* (Baltimore, 1979) is also considerable.

9   *Metropolitan Sanitary Commission: First Report* PP 1847–8:XXXII:48.

10  For a seminal discussion see Mary Douglas (1966).

11  *Report of the General Board of Health on the Epidemic Cholera of 1848 and 1849* PP 1850:XXI:543.

12  *Metropolitan Sanitary Commission*, 9.

13  *Report of the General Board of Health on Epidemic Cholera* (1850) Appendix B, 21.

14  *Cholera Report* (1848) lxi–lxvi.

15  *Report of the Medical Council* (1854) 6.

16  *Ibid*, 7.

17  *Ibid*, 116.

18  *Report of the Committee for Scientific Enquiries* (1854) 49. This phase has been meticulously reconstructed and assessed by Pelling, *Cholera, Fever and English Medicine 1825–1865* (Oxford, 1978) 203–50.

19  *Committee for Scientific Enquiries*, 51.

20  *Seventeenth Annual Report of the Registrar-General*, 99.

21  John Snow, *On Cholera*, edited by Wade Hampton Frost (New York, 1936) 124.

22  *Ibid*, 110.

23  *Select Committee on Public Health and Nuisances Removal Bill* PP 1854–5: XIII:Q 150.

24  Cited in J K Crellin, 'The Dawn of the Germ Theory: Particles, Infection and Biology' in F N L Poynter (ed), *Medicine and Science in the 1860's* (1968) 67.

25  'Report of Captain Tyler to the Board of Trade in regard to the East London Waterworks Company: PP 1867:LVIII:444.

26   W P G Woodforde in *Report of the MOH: Poplar* (1867) 35. See also his comments in *Cholera Report* (1866) 273.

27   *Cholera Report* (1866) 265. Evidence of Lionel J Beale, medical officer for St Martin-in-the-Fields.

28   *Report of the MOH: Bethnal Green* (1866) 20.

29   *Cholera Report* (1866) 263. Evidence of F Godrich.

30   *Ibid*, 278. Evidence of Henry N Pink.

31   'Special Report by Thomas Orton, MOH, Limehouse, on the Cholera Epidemic of 1866', 4.

32   *Report of the MOH: Limehouse, 1867 with Supplementary and Conclusive Remarks on the Cholera Epidemic in East London*, 10.

33   See, for example, the comments in *Eleventh Report of the MOH: Mile End Old Town*, 17; *Rivers Pollution Commission: Second Report: River Lea* PP 1867:XXXIII:Q 3220, evidence of James Knight, surveyor to the vestry of Mile End Old Town; and *Cholera Report* (1866) 263, evidence of Frederick J Burge, medical officer for Fulham.

34   This position is clearly articulated by J J Rygate in *Eleventh Report of the MOH: St George-in-the-East*, 26–7.

35   *Cholera Report* (1866) 191.

36   *Ibid*, 190.

37   *The Times* 2 August 1866.

38   *Ibid*.

39   *Cholera Report* (1866) 195.

40   *Ibid*, 196.

41   *Ibid*, 229.

42   *The Lancet* 11 August 1866.

43   *Cholera Report* (1866) 227.

44   'Captain Tyler's Report' (1867) 460.

45   *SC East London Water Bills* (1867) 363.

46   *RC on Water Supply* (1868) Q 3906.

47   'Captain Tyler's Report' (1867) 458–9.

48   *Rivers Pollution Commission, Second Report* (1867) Q 203.

49   *SC East London Water Bills* (1867) 662–4.

50   'Captain Tyler's Report', 448.

51   *Ibid*, 460.

52   *Rivers Pollution Commission, Second Report*, xxi.

53   *The Lancet* 2 November 1867.

54   *SC East London Water Bills*, xiii.

55   'Copy of the Correspondence between the Board of Trade and the East London Waterworks Company with Reference to Captain Tyler's Report on the Water Supplied by the Company' PP 1867:LVIII:481–93.

56   *Ibid*, 486.

57   'Robert G. W. Herbert to the Secretary of the East London Waterworks Company' PP 1867:LVIII:493.

58   *Cholera Report* (1866) xliv.

59   *British Medical Journal* 27 April 1867.

60   *Ibid* 2 November 1867.

61 *The Lancet* 2 November 1867.

62 *RC on Water Supply* (1868) Q 7127.

63 'Mr J. Netten Radcliffe on Cholera in London, and especially in the Eastern Districts', *Ninth Report of the Medical Officer of the Privy Council* PP 1867:XXXVII:Appendix 7, 368.

64 Lambert, *Sir John Simon 1816–1904 and English Social Administration* (1963) part IV.

65 'Radcliffe on Cholera', 331.

66 *Twelfth Annual Report of the Medical Officer of the Privy Council* PP 1870:XXXVIII:611.

67 *Cholera Report* (1866) 231–2.

68 But see J M Eyler, 'William Farr on the Cholera: The Sanitarian's Disease Theory and the Statistician's Method', *J. Hist. Med.* **28** (1973) 79–100.

# 5 Diarrhoea

Among the water-transmitted and water-related infections which are examined in this study, diarrhoea, and more particularly infant summer diarrhoea, is the most complex. Informed observers in the 1840s claimed that it was a 'new' disease which should not be confused either with dysentery (the venerable 'bloody flux'), ill-defined convulsions, long established as a cause of death during infancy and childhood, or cholera. But many medical men were slow to accept these differentiations or to acknowledge that, while adults might experience great suffering as a result of the illness, it was infants who were most gravely at risk. This discrepancy between 'advanced' medical thought and a majority of doctors led to repeated and at times confusing changes and refinements in the way in which diarrhoea was dealt with in the *Reports* of the Registrar-General—and there is an attempt to clarify these nomenclatures in the first part of this chapter.

In the mid-1850s pioneering epidemiological surveys by, among others, William Budd and William Ord, confirmed the irrefutably age-specific nature of the disease. In terms of possible environmental causation, however, there was massive and continuing uncertainty. Between the 1850s and the mid-1860s pollution of the Thames and the water derived from it were believed to play a crucial role in the dissemination of the infection. But, when by the early 1870s it could be seen that neither the partial repurification of the river nor the construction of the main drainage system had led to a significant reduction in mortality from infant diarrhoea, other hypotheses, to which the role of water became increasingly peripheral, began to be generated. And during the first great age of systematic investigation into the determinants of obstinately high levels of infant mortality, bacteriological research by Edward Ballard and others seemed to confirm that the infection was not in fact transmitted by a water-borne micro-organism.

Between the 1880s and the beginning of the First World War there were numerous national and regional assessments of the bacteriological status and environmental causes of diarrhoea. The most impressive of these studies were now multicausal rather than monocausal in character and pointed increasingly to the significance of domestic arrangements within individual dwellings. Within this new analytical framework, the presence or absence of a water-closet received increased emphasis. So, also, did the adequacy or otherwise of facilities for the cleansing of infants' eating and drinking utensils and the washing of clothes. Water, in other words, had been reintegrated within a more sophisticated set of explanatory variables. But the explicitly bacteriological conundrum was resolved neither by contemporary medical men nor, later, by social or demographic historians. And it is for this reason that the chapter concludes with an outline of some of the more credible speculations on the micro-organic identity of this elusive disease.

Diarrhoeal infection had been recorded in the capital, particularly during summer, for many years: but it evidently became much more severe during the 1840s. 'Diarrhoea', William Farr commented after the cholera of 1853–4 'is often a symptom of well-marked diseases; but this diarrhoea which always prevails in hot weather, and has been so common since the year 1846 is evidently a variety of cholera, proving chiefly fatal to young children and to old people, who do not so commonly exhibit spasms of cholera, but who have nearly all the other symptoms.'[1] Infants who were afflicted by the condition often suffered a period of 'dietetic injury' before the onset of the infection proper—with unsuitable, inadequate or, sometimes, over-large quantities of adult food causing damage to the digestive system.[2] When a household became infected adults would usually experience nothing more serious than nausea and an upset stomach: infants, though, would be ill for a period of between four days and a fortnight, and, in fatal cases, they would become gradually more emaciated and less able to eat. The terminal event usually took the form of a convulsion, and death was hastened, as in cholera, by extreme dehydration.[3]

Medical men throughout the nineteenth century were deeply puzzled by the pathological identity of diarrhoea, and this is reflected in the changing annual categories deployed by the Registrar-General between the 1850s and the late Edwardian period. These nosologies, which are relevant both to the quantitative parameters of the disease during our period, as well as to notions of the dominant mode of spread, are set out in table 5.1. Figure 5.1 compares national and metropolitan rates of mortality over six decades, and provides a broadly accurate guide to trends and fluctuations. It should be noted, however, that dysentery, which is included in each of the listings in table 5.1, was of no more than

**Table 5.1** Changing nomenclatures: diarrhoea, dysentery, cholera 1851–1910. Sources: *Annual Reports* and *Supplements* of the Registrar-General.

| | |
|---|---|
| 1851–60 | Diarrhoea, dysentery and cholera. |
| 1861–70 | Diarrhoea and dysentery (separated) and cholera. |
| 1871–80 | As for 1861–70. |
| 1881–90 | Diarrhoea and dysentery (not separated) and cholera. |
| 1891–1900 | Diarrhoea and dysentery (not separated) For those aged five and under diarrhoea was subdivided into enteritis; gastroenteritis; and gastritis. Cholera was enumerated separately. |
| 1900–10 | Diarrhoea and dysentery (separated). For those aged five and under diarrhoea was divided into diarrhoea due to food; infective enteritis and epidemic diarrhoea; diarrhoea (not otherwise defined); enteritis; and gastroenteritis. Cholera was enumerated separately. |

minimal statistical significance after the 1850s. (An analysis of age-specific mortality suggests that the condition probably accounted for fewer than a hundred deaths a year in London in the later nineteenth century.) Figure 5.1 shows that, compared with most other infectious diseases during this period, diarrhoea declined slowly and unevenly both nationally and in the capital. Indeed, during two decades 1861–70 (in London) and 1891–1900 (both in London and nationally) the downward trend was reversed. Diarrhoea, in other words, continued to be a massively resilient and destructive disease in nineteenth-century Britain. It accounted for over a thousand, mainly infant lives in London in 1850, over two thousand in 1875, and just under three thousand in 1900. During the traumatic epidemic of 1911 it killed over four and a half thousand—a death toll comparable with that attributable to cholera during the final metropolitan outbreak in 1866.

When we turn to an analysis of age-specific patterns during this period, it is clear from the upper curve in figure 5.2 that the infection gradually claimed more and more infant lives, with this group constituting approximately 80 per cent of total mortality by the beginning of the twentieth century. (No explanation has been found for the aberrant downward fluctuation in 1903.) The lower curve of figure 5.2 indicates deaths from diarrhoea as a percentage of aggregate annual infant mortality in London. Depending upon the severity of the disease in any given year, mortality from diarrhoea as a proportion of total infant mortality between 1840 and 1910 ranged from five to twenty per cent[4]: this may in fact be an underestimate since large numbers of deaths from

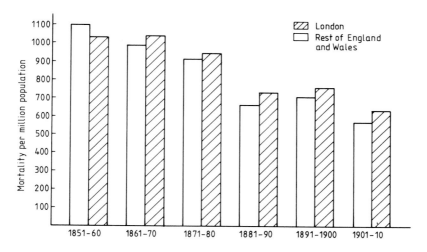

**Figure 5.1** Mortality from diarrhoea per million population, London and the rest of England and Wales 1851–1910. Sources: *Annual Reports* and *Supplements* of the Registrar-General.

**Figure 5.2** Mortality from diarrhoea among the 0–1 age group as a percentage of mortality from diarrhoea at all ages (top curve) and as a percentage of aggregate infant mortality (bottom curve), London 1850–1910. Source: *Annual Reports* of the Registrar-General.

convulsions associated with the condition continued to be attributed to the largest single category—'other causes'.

By the final third of the nineteenth century London was probably less severely afflicted by summer diarrhoea than other towns which experienced comparable rates of infant mortality. This was demonstrated by the findings of the epidemiologist W E Buck who devised a 'diarrhoea scale' for the years 1874–83.[5] Buck's calculations were based on the realistic assumption that, since the third and warmest quarter of the year always coincided with the highest rates of mortality from the disease, deaths during that period should be multiplied by four and expressed per thousand of population. Relativities remained broadly favourable until the outbreak of the First World War and beyond but as continuingly high levels of mortality and, traumatically, the epidemic of 1911, were to demonstrate, there was little room for complacency.

**Table 5.2**  Mortality from diarrhoea 1874–83 per 1000 population in selected urban areas. Source: W E Buck. 'On Infantile Diarrhoea', *Trans. San. Inst. GB* **vii** (1885–6) 87.

| | |
|---|---|
| Leicester | 7.1 |
| Hull | 5.0 |
| Birmingham | 4.2 |
| Norwich | 3.4 |
| London | 2.4 |
| Bristol | 1.7 |

When we examine district death rates from the disease, it is useful to concentrate on three groupings—those districts which recorded higher than average mortality for the period as a whole; those which registered higher than average rates during those decades (1861–70 and 1891–1900) when the rate for the metropolitan region rose following a decline during the preceding 10 years; and, finally, those 'backward' districts which displayed an increase in deaths from diarrhoea in the decade 1901–10 when mortality from the infection finally began to stabilise. This material is presented in figure 5.3.

Those districts which experienced a death rate from diarrhoea of 100 or more per 100 000 population throughout the period were Shoreditch, St George-in-the-East and Stepney. The second category consisted of Holborn and Islington for 1861–70 and Mile End, Poplar and St Saviour, Southwark for 1891–1900. Finally, the 'backward districts' between 1901 and 1910 were Holborn, City, Whitechapel and Poplar.

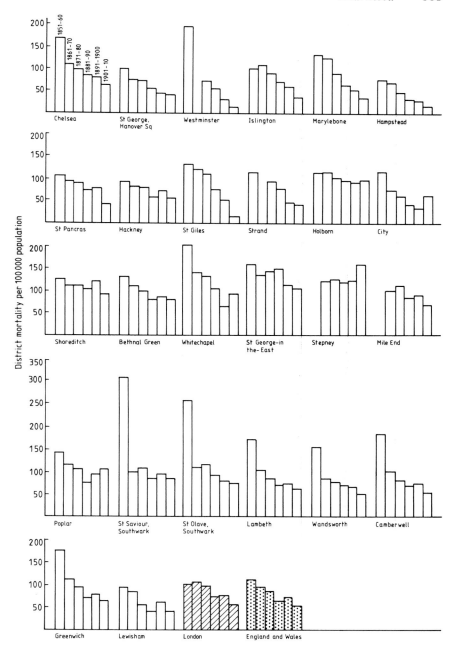

**Figure 5.3**  District mortality from diarrhoea per 100 000 population, London 1851–1910. Sources: *Annual Reports* and *Supplements* of the Registrar-General. The omission of columns in a small number of cases indicates that the data are unreliable.

In terms of poverty, social structure, population density and environment the first category was homogeneous—indeed, in 1894 each of
these East End districts was designated as being characterised by above
'average London poverty' according to a scale based on Booth's survey
technique.[6] As a result of inner-city overcrowding Holborn had become
highly vulnerable by the 1860s but adverse conditions were altogether
less evident in Islington.[7] In the case of those districts which were
particularly at risk between 1891 and 1900, we need to distinguish
between levels of poverty and complex environmental conditions
intimately related to poverty. Shoreditch, Mile End and Poplar were
each located at the very heart of the East End but, in socio-economic
terms, they were far from identical—by the 1890s, for example, Mile
End was thought to be distinctly more well-to-do than either Shoreditch
or Poplar.[8] The addition of St Saviour, Southwark to this grouping in
1891–1900 was hardly unexpected since its record, both in terms of
overcrowding and general levels of health, had long given cause for
concern. Between 1901 and 1910 Holborn, the City and Whitechapel
were all adversely affected by inner-city depopulation and dishousing
with the better-off members of these communities moving to the
suburbs: in these areas, also, late-nineteenth-century sweated and
service occupations had not yet been replaced by higher paid and
healthier work.[9]

One final primarily quantitative and epidemiological issue, namely
the relationship between cholera and diarrhoea, requires clarification.
A number of doctors and public health officials were convinced that the
poor, and those who treated them, often deliberately exaggerated the
incidence of diarrhoea rather than admit that their district might
already be in the grip of cholera. It was also frequently insinuated that
this minimisation of cholera allowed local oligarchies to delay putting
preventive regulations into effect. At the height of the crisis in London
in September 1854, *The Lancet* demanded 'Why, we would ask, is
Cholera . . . designated by two titles or names?'[10] In the same year the
Committee for Scientific Inquiries commented both on the arbitrary
nature of diagnosis during such epidemics and more especially on
differences in medical certification among those dying at home rather
than in hospital. 'In private practice', the committee claimed,
'especially in districts where the disease was not very rife there was a
disposition to give the less formidable name of "diarrhoea" even to
cases which had the features of cholera distinctly marked: whilst in
hospitals, a more strictly pathological view of the matter being taken,
the fatal cases were denominated deaths from cholera.'[11] Metropolitan
Poor Law officials were also believed to be reluctant to admit that
cholera had breached the walls of the workhouse. Of the 842 people who
had died from cholera or diarrhoea in London hospitals in 1853–4, 800

were deemed to have been victims of cholera; of 1325 dying from both diseases in workhouses, more than a quarter were claimed to have been struck down by diarrhoea.[12] Yet informed contemporaries were undoubtedly aware that deaths attributed to diarrhoea during cholera years were concentrated, as they were during non-cholera years, predominantly among those under the age of one.[13] This suggested that diagnosis may have been less fallible than critics implied.

It seems probable, in fact, that during cholera epidemics medical men were confronted by three distinct categories of patients: those suffering from cholera in an acute form, from which they would be unlikely to recover; those displaying the symptoms of 'choleraic diarrhoea' or cholera at a less advanced stage, which they might well survive; and those—mainly the very young—who had contracted the diarrhoea which was endemic and sometimes seriously epidemic in London and every other urban area during the summer and early autumn.[14] This may have been the pattern observed by Thomas Sarvis, medical officer for Bethnal Green during the cholera epidemic of 1866, among those seeking treatment at his dispensary. Sarvis estimated that 366 were suffering from cholera, 822 from 'choleraic diarrhoea' and no fewer than 16 345 from diarrhoea.[15] There are indications here that inhabitants of the East End were seeking treatment at all ages for conditions which they would have been willing to leave untreated in more normal times, when medical provision was in any case less widely available and less widely publicised. Further light is thrown on this issue by the comments of a doctor who worked in Hackney and Shoreditch and who reported that when he asked in the aftermath of the cholera of 1849 'if any person were ill, the almost inevitable answer was "No, but my husband or child has got a very bad bowel complaint".' 'One reason for this apathy', the doctor went on, 'consists in the belief of the poor that everything of the kind will "work itself off": this belief probably arising from the frequency of diarrhoea among them.'[16]

There is, in fact, a moderately reliable method of deciding whether the diarrhoea diagnosed as a cause of death during cholera epidemics was in fact the disease 'prevailing annually and at all times in the country' or merely a euphemism for cholera proper. Figure 5.4 shows annual mortality from diarrhoea in London between 1845 and 1870, a period which encompassed both the emergence of the disease as a recognised entity and the cholera epidemics of 1848–9, 1853–4 and 1866. It will be seen that it was only in 1849 that mortality from the infection rose significantly above the decennial average: and that, neither in 1853, 1854 nor 1866, did deaths from this cause move dramatically upwards. It is, in fact, noticeable that two of the three peaks—1857 and 1859—coincided with non-cholera years and that mortality from diarrhoea in 1866 was below the high levels prevailing in

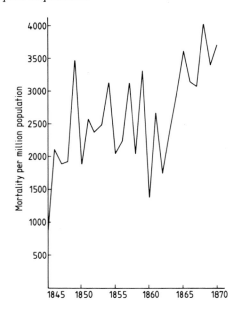

**Figure 5.4** Mortality from diarrhoea per million population, London 1845–70. Source: *Annual Reports* of the Registrar-General.

the period 1865–70. We may therefore conclude that estimates of the 'balance' between the two diseases arrived at by the Registrar-General were moderately accurate and that it was only in 1849 that a 'transfer' from diarrhoea to cholera should probably have been made. (The exceptionally high rates in Southwark during the single decade 1851–60 may also indicate widespread misdiagnosis in that area during the cholera epidemic of 1853–4.)

The ground has now been prepared for an examination of changing ideas on the nature and causation of diarrhoea and, more specifically, of notions about how the disease may have been related to the consumption and use of unsafe water. The wide-ranging and important debate which took place during the 1850s about the role of contaminated supplies in the transmission of cholera was paralleled by a comparable flurry of activity concerned with possible associations between polluted water and diarrhoea.

Major revisionist surveys were undertaken by William Budd, William Ord, Edward Greenhow and John Snow. The conventional wisdom during the 1840s and early 1850s was that an unprecedentedly polluted Thames had propagated intolerably high levels of 'fever'. Infection, according to this model, had been transmitted, not as a result of the actual consumption of foul water, but atmospherically, with vapours rising insidiously upwards from the surface of the river. In order to counteract this orthodoxy in relation to diarrhoea, Budd reworked readily available material to emphasise the importance of the faecal–oral route as a mode of spread. Writing about the environmental crisis of 1858, he concluded that 'the returns of sickness and mortality were made up, and, strange to relate, the result showed not only a death-rate below the average but, as the leading particularity of the season, a remarkable diminution in the prevalence of fever, diarrhoea and other forms of disease commonly ascribed to putrid emanations'.[17] He added that death rates from diarrhoea, choleraic diarrhoea and dysentery were 26.3 per cent lower in the summer of 1858 than they had been over a comparable period during the three preceding years, and 73 per cent lower than in 1857.[18]

Ord's inquiry on behalf of the Medical Office of the Privy Council was more detailed. First, he described the symptoms which riverside workers had displayed during the Thames crisis of 1858. 'The attacks', he reported, 'began with langour and depression, followed by nausea and headache, particularly severe in the region of the forehead and temples by aching of the eye-balls, and by a red and swollen state of the throat (erythema). To these were often added giddiness, temporary loss or impairment of sight, the presence of black spots before the eyes and even utter mental confusion.' Whatever this exceedingly unpleasant condition might be, Ord explained, it ought not to be diagnosed as diarrhoea.[19] Next, following Budd's lead, Ord showed that deaths from diarrhoea among riverside workers had not risen nearly so dramatically in 1858, when the river had been in a disgraceful condition, as in 1859 when it had been less noxious and more thoroughly deodorised.[20] He also demonstrated that, in 1859, when, during a single week in July, deaths from diarrhoea had risen as high as 412, the great majority had been among the capital's infant population.[21] This threw the Thames-borne theory into further disarray since it raised the question of why it was that individuals far distant from, and seemingly meteorologically insulated from the foul river-course, should have been attacked by epidemic vapours: and why the youngest age group of all, rather than a more random sample of the population, should have been so severely affected.

Making use of a comparison of death rates in extra-metropolitan Surrey and the 'rest of London' John Snow, like Budd, re-evaluated existing statistical material concerned with 'fever' and diarrhoea.[22] This was an analysis which was tacitly antagonistic to miasmatic dogma, and its influence, taken in conjunction with Budd and Ord's contributions, ensured that by the early 1860s medical officers of health were beginning to acknowledge that diarrhoea seemed to exert its deadly influence in districts far removed from polluted river-courses and foully-impregnated marshes and upon groups, particularly the very young, who seemed often to be protected from the worst of the 'epidemic atmosphere'.[23] Important aspects of existing orthodoxy as to the nature of diarrhoea were therefore called into question. The disease appeared to possess a dynamic of its own and one which was quite different from the diarrhoea associated with cholera: it was much more threatening to the very young than to adults; and it could no longer be assumed to be disseminated miasmatically—whether that doctrine were interpreted in its most formal sense, or in terms of a polluted river interacting with an 'epidemic atmosphere'.

Finally, in a classic survey in 1860 Edward Greenhow confirmed that fatal diarrhoea should be primarily construed as a disease of childhood and infancy. Although he was committed to a still broadly miasmatic system of thought, and was thus convinced of the causal significance of 'foul air' inhaled from privies constructed too near to overcrowded dwellings, Greenhow also gave prominence to unsafe water. 'As far as I know', he asserted, 'it is a practical certainty, that, *in the districts which suffer the high diarrhoeal death rates, the population either breathes or drinks a large amount of putrefying animal refuse.*'[24]

An association between polluted water and diarrhoea was also now recognised by less prestigious members of the scientific and medical communities. By 1850 it was widely believed that to drink directly from the Thames or Lea was to risk serious diarrhoea or dysentery.[25] In the poorest areas of the capital children who were feverish, and all the more likely to want to consume large quantities of water, were often prevented from doing so by their mothers. 'In cases of sickness and fever', a London doctor lamented, 'when the natural inclination is for water a child will ask for water, and the mother will say: "Is it right for me to give it? I think that it must be injurious . . .".'[26] Unfiltered supplies, derived directly from the Thames at Millbank Prison in the 1850s, were believed to have led to high incidence of diarrhoea[27]: and in the later 1850s and early 1860s in Paddington, St James, Westminster and Fulham, consumption of water from surface wells was blamed for outbreaks of the infection.[28] In St James, Westminster in 1859, 'the tendency to diarrhoea' was reported to have 'increased in the summer season of the year by the drinking of unfiltered and surface well-water, which, after standing a little time, becomes putrescent, and is capable of producing the same state of change in food with which it is mixed'.[29] By the late 1850s public health officials believed that a fall in the incidence of diarrhoea might be related to reduced reliance on well water and increased consumption of safer company water.[30] A decade later this mode of analysis had been reversed, with abnormal prevalence of the infection being adduced as evidence that the companies were still providing intermittently dubious supplies[31] and by the early 1880s medical officers were regularly identifying unsafe company water by way of widespread outbreaks of diarrhoea.[32] During the acute water shortage in east London in 1895 cases multiplied dramatically[33]; and in 1897, the first unsuccessful effort was made to deploy microbiological evidence to pin down a small-scale outbreak of the disease.[34] But much of this observation and intervention was directed towards non-fatal cases among adults and it was only in the 1880s that a new generation of epidemiologists and medical scientists extended the pioneering work of Ord and Greenhow and brought their expertise to bear on the specifically infant form of the infection. It was clear that more effective sewage disposal techniques and a safer water supply had not reduced exceptionally heavy mortality from the infection among the youngest age groups in urban areas. And, as Edward Ballard concluded in a classic study in 1889, there were now good grounds for believing that the causal pathogen was only rarely transmitted in unsafe water. 'Undefined polluted conditions of Drinking Water', he stated, 'have from time to time appeared to give rise to epidemic diarrhoea, and this irrespective of the time of year; but, so far as I have been able to ascertain, water has little or nothing to do in the way of direct causation with the diarrhoeal mortality which occurs annually in this country in the summer.'[35] Surveying the history of the disease in London from the vantage point of the 1890s, Sims Woodhead was also sceptical of ever being able to isolate a specific diarrhoea micro-organism in water.[36]

Specialist modes of thought as to how diarrhoea was transmitted appeared to have turned full circle. At the beginning of the period under discussion, medical men had acknowledged the importance of river and water pollution, but only, as with cholera, within a miasmatic or Thames-borne theory of infection which claimed that foul water exacerbated an already pernicious 'epidemic atmosphere'. During the 1850s, and more particularly during and immediately after the environmental crisis of 1858, this orthodoxy was undermined by investigators, notably Budd and Snow, who had already propounded anti-miasmatic theories in relation to cholera and typhoid and who now turned their attention to diarrhoea. Others, particularly Ord and Greenhow, were less concerned with the aetiological nature of the condition than the extent to which it had now revealed itself as an infection of infancy and childhood. During the 1860s these innovations gained wider currency among public health officials, as also did the role that epidemics of diarrhoea could play in pointing to unequivocally substandard well water. By the 1870s and 1880s, as water and sewage disposal systems underwent substantial improvement, it became clear that such measures had not reduced the appalling death rate still attributable to summer diarrhoea. And it was for this reason that epidemiologists, a majority of whom were now committed to one or other version of the germ theory, became convinced that the infection was not in fact usually carried in polluted water.

Running alongside these modes of explaining diarrhoeal infection, there were other, and at times divergent, systems of thought which lend support to the idea that scientific paradigms are invariably accompanied by 'non-theoretical' popular beliefs about the human body, health and disease.[37] The dominant lay attitude towards diarrhoea during the period when the miasmatic doctrine held sway, and was then slowly undermined by the innovations of the late 1850s and the 1860s, was a simple and commonsensical one—foul river and drinking water were likely to make you sick and to give you a painful and 'heavy' stomach. Even though it might fail to confront the question of causality, this deeply embedded oral and experiential tradition probably constituted as good a guide to the protection of health as medically authenticated orthodoxy.

By the 1890s attention among medical men and scientific investigators had become more heavily concentrated on the internal environment of the individual dwelling. Dirty babies' bottles, which were too often dipped into a saucepan of already tainted water, had in fact been identified as a possible source of infection in London as early as the 1870s.[38] Connections were now also being made between diarrhoea and bottle feeding at a time when it was widely believed that the latter practice was rapidly replacing the more 'natural' procedure of breast feeding.[39] Unsafe and contaminated food, particularly condensed milk and other tinned products, were also subjected to scrutiny. And each of these explanations was linked to arguments which emphasised that women's work allowed too little time to be spent on domestic hygiene and subjected the very young to the inexpert care of grandparents and baby minders.[40] Supplies of water for the washing of bodies and clothes were still inadequate in the poorest sections of the capital: and wide differentials in income ensured that the working class had no option but to buy inferior, and grossly adulterated, food and milk.[41] District and class differences in infant mortality and mortality from infant diarrhoea were, in other words, inseparable from the sociopolitical and geographical structure of the late-nineteenth- and early-twentieth-century city. It might be analytically useful to distinguish between the impacts of the internal and external environments in the precipitation of the disease, and to record and explain differing rates of change within each of these spheres; but access to healthy surroundings and freedom from infection and death among the infant population was fundamentally determined by an ability to buy oneself and one's family out and away from the most squalid and overcrowded areas.

Epidemiologists and scientists in the early twentieth century continued to search for ecologically and environmentally convincing explanations of infant diarrhoea, and the role of the fly as vector now became a major focus for research. The best-known investigator in this field was the American Charles V Chapin, but he was in due course forced to disown some of the more disingenuous monocausal accounts which were derived from his work and publicised by over-zealous public health reformers.[42] Elements of a more genuinely multicausal analysis were assembled in 1899 by Arthur Newsholme, who isolated six variables—urbanisation, poverty, meteorological and climatic conditions, poor 'soil', the unsatisfactory cleansing of streets and the absence of a modern water-carried system of sewage disposal—as determinants of the disease.[43] Newsholme's contention that adequate provision of water-closets was an essential precondition for a reduction in mortality from infant diarrhoea was a convincing one. Thus relatively affluent districts such as St James, Westminster, Chelsea and St George, Hanover Square, were all well provided with water-closets throughout our period and recorded low rates of diarrhoeal infection.[44] But in two inner-city districts, Holborn and Poplar, which were frequently ravaged by the disease, only 101 and 251 closets respectively were installed at public expense between 1856 and 1870.[45] Evidence for the later part of the century is difficult to evaluate but the fact that it was only during the 1890s that more than 50 per cent of the total population living in the region bounded to the east by Poplar, to the north by Hackney, and to the west by St Pancras, had access to constant supply, argues against widespread adoption in the poorest areas of the capital.[46]

There were, then, stark class- and income-related factors militating against a rapid reduction in mortality from the 'summer infection' in the early twentieth century. But what was undeniable was that important aspects of environmental thinking having a close bearing on the control of diarrhoea had been transformed. In 1858 the medical officer for Holborn had pondered whether, in the light of public outcry during the crisis of that year, he should instruct his parish officers to postpone the systematic destruction and filling in of cesspools. But, having examined the health statistics, and noted the apparent reduction in mortality from every category of infection, he concluded that the purity of the Thames should not be preserved at the expense of lives lost in Holborn.[47] Similarly, in 1860, the medical officer for Shoreditch wrote: 'It was thought by many persons of influence to be better to live in the midst of over-flowing cesspools than to add to the defilement of the river. Happily this idea never prevailed in Shoreditch.'[48]

Forty years later, with the sewage of the capital being transported to Crossness and Barking, such battles had long since been fought and decided. It was, in other words, now feasible to advocate a wholly water-carried sewage system without having to condone pollution of the inner-city river and the higher mortality from cholera and typhoid which such a policy had implied in the period before the water companies secured safer sources of supply. By the early twentieth century, then, a more reliable water supply combined with a moderately efficient main drainage system and larger, though not yet large enough, numbers of water-closets, had made a major contribution to the reduction in mortality from the 'filth diseases' and notably cholera and typhoid both in London and nationally.[49] But because of its bacteriological resilience and concentration among the most sensitive age group of all, the infant population, diarrhoea had reacted unsatisfactorily to this programme of urban reform. In the absence of radical improvements to public utilities and services, the death rate from the infection would no doubt have been even higher between 1840 and 1910 than in fact it was. But until quantities of water supplied to the poorer districts could be increased up to a point at which it would be possible for every family to have access to a water-closet, and to be able to wash babies' clothes and eating and drinking utensils more regularly and thoroughly, further amelioration would be delayed. And changes of this type would only be achieved when metropolitan levels of poverty had been comprehensively reduced, allowing a substantially larger proportion of the population to move to less crowded and better constructed accommodation.[50] This would ensure, also, that poorer members of the community would have access to more easily laundered sheets and clothes, disinfectants and, if and when serious infant and childhood illness did break out, to medical and nursing advice. It was this long-term transformation, then, rather than any specific epidemiological initiative, or the adoption by working people of the advice too earnestly thrust upon them by 'sanitary visitors' which may finally have led to the elimination of the disease.[51] As for large-scale supplies of clean drinking water, these probably constituted a necessary but not a sufficient condition for the eradication of the deadly summer infection. What, finally, may be concluded about the bacteriological status of diarrhoea during this period? It is probably unwise, firstly to over-stress the similarities between the nineteenth-century infection and the modern disease which kills about five million infants and young children in Afro-Asia annually.[52] Nor should summer diarrhoea be too readily equated with the gastro-enteritis which periodically traumatises modern maternity and mother-and-baby units. The suddenness of decline after about 1911 may indicate a form of 'flame-out' comparable with, but less absolute than, that which may have contributed to the elimination of bubonic plague in western Europe between the seventeenth and later eighteenth centuries.[53] A further hypothesis is that the disease was spread by an unstable virus which had become substantially less dangerous by the beginning of the second decade of the twentieth century.[54] Or—a final possibility—the infection may have flourished and waned in rhythm with the unpredictable *Proteus morganii*, a microorganism frequently found in human faeces.[55] None of these speculations can be anything other than provisional but existing evidence, both in the form of case histories and material on the course and periodicity of epidemics during the later nineteenth century, is certainly rich enough to warrant further investigation.

1    'The Cholera Epidemic of 1853–4' in *Seventeenth Annual Report of the Registrar-General*, 75.
2    Leonard Parsons and Seymour Barling (eds), *Diseases of Infancy and Childhood* Vol 1 (Oxford, 1954) 591 and H S Banks, *The Common Infectious Diseases* (1949) 326.
3    Hugh R Jones, 'The Perils and Protection of Infant Life', *J. Stat. Soc.* **57** (1894) 21.
4    Nationally, a peaking of the disease was noted in the 1890s by Arthur Newsholme, *The Elements of Vital Statistics* (1923) 425.
5    W E Buck, 'On Infantile Diarrhoea', *Trans. San. Inst. GB* **vii** (1885–6) 87.
6    *RC Water Supply Metropolis* (1893) Appendix C Q18, 225.
7    Gareth Stedman Jones, *Outcast London: A Study in the Relationship between Classes in Victorian Society* (Oxford, 1971) 175 and 232.
8    *RC Water Supply Metropolis* (1893) Appendix C Q18, 225.
9    For Whitechapel and a succinct account of the region as a whole see P G Hall, *The Industries of London since 1861* (1962) 64.
10   *The Lancet*, 9 September 1854.
11   *Committee for Scientific Enquiries* (1854) 66.
12   *Ibid*, 65.
13   *Ibid*.
14   Relevant clinical aspects of cholera are discussed by S N De, *Cholera: Its Pathology and Pathogenesis* (Edinburgh and London, 1961) 1–3.
15   *Report of the MOH: Bethnal Green* (1866) 7.
16   *Report of the General Board of Health on Epidemic Cholera* (1851) 21. Evidence of Dr Gavin.
17   William Budd, *On Typhoid Fever* (1873) 150.
18   *Ibid*, 151.
19   William Ord, 'Proceedings in Reference to the London Sewage Nuisance' *Second Report of the Medical Officer of the Privy Council* PP 1860:XXIX:256.

20    *Ibid*, 258.
21    *Ibid*.
22    John Snow, *Medical Times and Gazette* 20 February 1858, 161–3 and 188–91.
23    See, for example, *Report of the MOH: Lambeth* (1860) 13.
24    E H Greenhow, 'Proceedings in Reference to the Diarrhoeal Districts of England', *Second Report of the Medical Officer of the Privy Council* (1860) 362.
25    *Report of the General Board of Health on the Supply of Water to the Metropolis* (1850) Qs 724–31. Evidence of Dr Gavin and Robert Bowie, surgeon.
26    *SC Metropolis Water Bill* (1851) Q 6139. Evidence of Dr John Challice.
27    *SC Public Health and Nuisances Removal Bill* (1854) Qs 169–70. Evidence of John Snow.
28    *Paddington: Vestry Report* (1859) 27–8 and *Report of the MOH: Fulham (1863) 13.*
29    *Report of the MOH: St James, Westminster* (1859) 14.
30    *Report of the MOH: Kensington* (1857) 31.
31    *Report of the MOH: Wandsworth* (1868) 20.
32    *Report of the MOH: Hackney* (1882) 18–20.
33    *Report of the MOH: Hackney* (1895) 46–9 and *Report of the MOH: Bethnal Green* (1895) 33.
34    *Report of the MOH: Stoke Newington* (1897) 29.
35    'Diarrhoea and Diphtheria' in *Supplement in Continuation of the Report of the Medical Officer of Health for 1887* PP 1889:XXXV:9.
36    *RC Water Supply Metropolis* (1893) Appendix C Q 17 501.
37    Here one might juxtapose the arguments put forward in relation to 'high science' by Thomas Kuhn, *The Structure of Scientific Revolutions* (Chicago, 1962) and the lay-specialist dichotomy which is central to Keith Thomas, *Man and the Natural World: Changing Attitudes in England 1500–1800* (1983).
38    Hugh R Jones as note 3 above.
39    *Report of the MOH: Kensington* (1871) 9–10.
40    E H Phelps Brown, *The Growth of British Industrial Relations* (1960), 40. See also Carol Dyhouse, 'Working Class Mothers and Infant Mortality in England, 1895–1914', *J. Soc. Hist.* **12** (1978–9) 248–67.
41    Robert Roberts, *The Classic Slum: Salford Life in the First Quarter of the Century* (Manchester, 1971) chapter 6.
42    James H Cassedy, *Charles V. Chapin and the Public Health Movement* (Cambridge, MA, 1962) 101–2. A H Gale, *Epidemic Diseases* (1959) 87 finds the fly theory 'plausible but unprovable' but Lambert, *Sir John Simon 1816–1904 and English Social Administration* (1963) 320 and 600, and Parsons and Barling (note 2 above) 587, are sympathetic.
43    The best synthesis was Arthur Newsholme, 'A Contribution to the Study of Epidemic Diarrhoea', *Public Health* **xii** (1899) 139–211.
44    *Metropolitan Sanitary and Street Improvements* PP 1872:XLIX:585 *passim*.
45    *Ibid*, 589, 629 and 634.
46    These calculations are taken from my article in R Woods and J Woodward (eds), *Urban Disease and Mortality in Nineteenth Century England* (1984) 111–12.
47    *Report of the MOH: Holborn* (1859) 28–9.

48    *Report of the MOH: Shoreditch* (1860) 16.
49    This is the general argument which is central to Thomas McKeown, *The Modern Rise of Population* (1976).
50    But this was a long-delayed process. For an authoritative account see Anthony S Wohl, *Endangered Lives: Public Health in Victorian Britain* (1983) chapter 11 and the same author's *The Eternal Slum: Housing and Social Policy in Victorian England* (1977).
51    The ideological underpinnings of 'home advice' in the earlier twentieth century have been extensively investigated by Anna Davin in 'Imperialism and Motherhood', *History Workshop* **5** (1978) 9–67.
52    But see F B Smith, *The People's Health 1830–1910* (1979) 85–104 and Anthony S Wohl (1983) (note 50 above) 22–5. On food patterns in the developing world Alan Berg, *The Nutrition Factor: Its Role in National Development* (Washington, 1973) is indispensable. There are also interesting cross-cultural implications in Khin-Maung-U *et al*, 'Effect on Clinical Outcome of Breast Feeding during Acute Diarrhoea', *Br. Med. J.* **290** (1985) 587–9. My thanks to Jane Martin for this reference.
53    See Andrew Appleby, 'The Disappearance of Plague: A Continuing Puzzle', *Econ. Hist. Rev.* **33** (1980) 161–73 and Paul Slack, 'The Disappearance of Plague: An Alternative View', *Econ. Hist. Rev.* **34** (1981) 469–76.
54    This is a view which has recurred in the writings of A H Gale. See his *Epidemic Diseases* (1959) 84 and 'A Century of Changes in the Mortality and Incidence of the Principal Infections of Childhood', *Arch. Dis. Childhood* **20** (1945) 14–15.
55    Joan Taylor, 'Infectious Infantile Enteritis, Yesterday and Today', *Proc. R. Soc. Med.* **63** (1970) 1294 and David Morley, *Paediatric Priorities in the Developing World* (1973), 179. On the possible role of insufficient quantities of water as a precipitant of infectious diarrhoea during infancy see Morley's comments at 170–94 and, particularly, 184.

# 6  Typhoid Fever

Typhoid was endemic in preindustrial society.[1] But as urban populations grew with increasing rapidity from the later eighteenth century onwards, and were supplied with water and food from more homogeneous sources, the disease claimed ever-larger numbers of victims. Most preindustrial and proto-industrial villages and small towns had been dependent upon wells and streams for their drinking water, and it was when these sources became polluted that typhoid struck, mainly during the autumn. But in the cities of the early and mid-nineteenth century the infection became an altogether more serious threat. Indeed, if the annual autumn death toll in London between about 1850 and 1870 is 'spread' evenly throughout the year it is probable that as many as four people a day perished from the disease. The death rate declined after 1870, but there can be no doubt that, among the predominantly water-transmitted infections under examination in this study, it was typhoid which accounted for the largest number of deaths in London between 1840 and 1910.

When an individual contracts the disease he or she becomes tired and weak, loses appetite and runs a very high temperature. Diarrhoea then sets in and in fatal cases abdominal haemorrhage attributable to diarrhoea is often the terminal event.[2] It was only in 1869 that typhoid was differentiated from the louse-borne infection, typhus, in the annual reports of the Registrar-General. Clinically, however, a number of medical men had made a distinction between the two conditions earlier in the nineteenth century. In 1826 P C A Louis in Paris had used the term 'typhoid' to describe the delirium which often accompanies the fever; and, subsequently, a fuller identification was made by the American William Wood Gerhard and confirmed by Jenner in 1849 and 1853. An alternative nomenclature, 'enteric', which was in due course adopted by the Registrar-General was introduced by Ritchie in 1846 to indicate the bowel symptoms which are the most important feature differentiating typhoid from typhus.[3]

William Budd's classic investigations consolidated the idea of a distinct and specific disease entity, most frequently transmitted by contaminated water and food.[4] Nevertheless, with typhoid, as with other 'fevers' in the nineteenth century, miasmatic doctrine, and variants on it, remained highly influential until a remarkably late date. Thus, in numerous editions of a standard teaching manual, Florence Nightingale was authoritatively quoted as believing that 'diseases begin, grow up, and pass into one another. . . . I have seen, for instance, with a little overcrowding, continued fever grow up; and, with a little more, typhoid fever; and, with a little more, typhus.'[5] According to this paradigm, which has already been discussed in relation to cholera, 'fevers' were believed to be generically interrelated, and their incidence determined by squalor, the 'vapours' generated by squalor, and the 'moral' attitude adopted (or not adopted) by vulnerable individuals. This conception of the nature of fever played an important role in ensuring that misdiagnosis of typhoid remained frequent until the infection became reportable under the Infectious Diseases Act in 1889. Indeed, statistics for the first full decade of notification suggest that in London, and nationally, typhoid may have been seriously under-registered between the time of its official separation from typhus and the early twentieth century.[6] It was only in the early 1900s that the Widal test allowed markedly fewer cases of the infection to avoid detection.[7] Greater accuracy had by then been achieved as a result of cooperation between medical officers, public analysts, fever and isolation hospitals and bacteriological laboratories.

In the first part of this chapter an attempt is made to disentangle typhoid from typhus in London between about 1840 and 1870 and to outline the quantitative parameters of the infection during the period, from the 1870s onwards, in which it was separately recorded by the Registrar-General. This is complemented by an examination of the role which medical men and other investigators ascribed to unsafe water and other media as modes of transmission. The chapter concludes with an evaluation of the ways in which this ubiquitous infection shaped the theory and practice of epidemiology in the later nineteenth and early twentieth centuries.

In figure 6.1 combined mortality from typhus and typhoid in London between 1840 and 1868 is compared with mortality for the rest of England and Wales. It can be seen that the 1840s were characterised by very high numbers of deaths from this group of infections. Available evidence, however, supports the view that it was typhus, particularly during the disastrous epidemics of 1838–42 and 1846–7, rather than

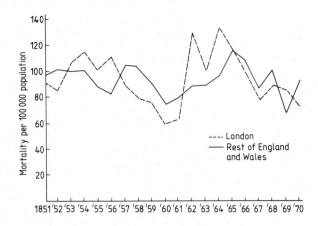

**Figure 6.1** Mortality from undifferentiated typhus and typhoid, 1851–70, per 100 000 population. Source: *Annual Reports* of the Registrar-General.

typhoid, which was dominant at this time.[8] For the 1850s, analysis of 18 000 cases admitted to the London Fever Hospital between 1848 and 1870, taken in conjunction with Charles Creighton's estimates, confirm that during this decade typhoid claimed a growing proportion of a slightly declining aggregate mortality from the two infections.[9] This is a trend which is substantiated by an unusually early and thorough report which sought to separate the diseases between 1855 and 1863 in the fever-prone district of Shoreditch (see table 6.1). Table 6.1 suggests that from the early 1860s onwards it was typhus which was once more claiming increasing numbers of victims; and between 1862 and 1870 the louse-borne infection returned in the form of a fierce and, as it transpired, final onslaught, centred on the poorest sections of the eastern, north-eastern and inner-city districts. There is, however, no indication of a corresponding decline in typhoid mortality during this epidemic— and data from the London Fever Hospital show that between 1863 and 1870 mortality from typhoid followed the upward movement of typhus, although at lower absolute levels.[10] For London, at least, the implication is that the 1860s were an exceptionally unhealthy and infection-ridden decade—a pessimistic epidemiological conclusion, relating both to poverty and the quality of life, which has yet to be confronted by historians involved in the standard of living debate.[11]

It is clear that the 'mix' between typhoid and typhus cannot, in the very nature of things, be uncovered in figure 6.1. In table 6.2, therefore, an attempt has been made to summarise the probable proportional balance between the two diseases during this period. From 1869 onwards it is possible to be more certain about the trend of mortality from typhoid in London. Figure 6.2 shows that metropolitan rates were generally lower than those for the rest of England and Wales throughout the period from 1870 to 1910.[12] During the early 1880s, however, there was a brief reversion to higher levels of mortality. Following continued decline in the mid-1880s, the rate remained sticky for the rest of the century before final stabilisation at unprecedentedly low levels by about 1905.

**Table 6.1** Death rates from fever per 10 000 population in Shoreditch, 1855–63. Sources: *Report of the MOH: Shoreditch* (1864) 17 and *Annual Reports* of the Registrar-General.

| | | |
|---|---|---|
| 1855 | 14.1 | |
| 1856 | 20.6 | |
| 1857 | 16.8 | Typhoid predominant |
| 1858 | 12.3 | |
| 1859 | 9.9 | |
| 1860 | 8.6 | |
| 1861 | 9.5 | |
| 1862 | 12.2 | Typhus predominant |
| 1863 | 11.1 | |

The process underlying the reduction in national typhoid mortality has been summarised as the 'sanitary revolution'.[13] But it has also been noted that the improvement was uneven and less dramatic than that which characterised the decline of typhus from the 1850s onwards.[14] Greenwood isolated three phases. He argued that between 1875 and 1885 the rapid fall was due mainly to a 'vigorous overhaul of drains and water'. Between 1885 and 1895 little change could be discovered and 'the number of utterly detestable water supplies was, relatively to the whole number of supplies, so small that not much impression upon the *national* rates could be made by impeaching water and drains: there was an epoch of stagnation'. After 1900 there was a further decline as attention was turned to the rapid hospitalisation of infected patients and the mechanism of the immune carrier.[15] Within this national context the great majority of those professionally concerned with public health believed that the reduction of the metropolitan typhoid rate during the last quarter of the nineteenth century gave cause for satisfaction.[16] Palmberg showed that throughout the period 1850–90 mortality from fever and, for the period for which there were adequate statistics, mortality from typhoid had been substantially lower in

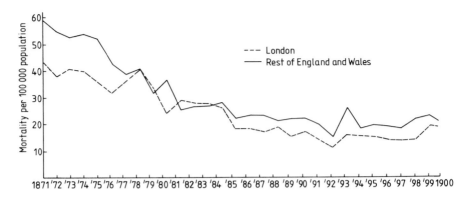

**Figure 6.2**  Typhoid mortality, 1871–1900, per 100 000 population. Source: *Annual Reports* of the Registrar-General.

**Table 6.2**  Typhus and typhoid in London, 1840–70. Sources: *Annual Reports* of the metropolitan medical officers of health and the Registrar-General. See also note (10) to this chapter.

| | |
|---|---|
| 1841–50 | High levels of mortality from both infections, but with the aggregate dominated by *typhus*, mainly as a result of the epidemics of 1840–2 and 1846–7. |
| 1851–60 | Substantially lower levels of aggregate mortality, but with *typhoid* claiming a higher proportion of total deaths from the two diseases. There was, however, an important *typhus* epidemic associated with the recession at the end of the Crimean War (1856–7). |
| 1861–70 | *Typhus* returned in a final onslaught between 1862 and 1870. This epidemic was, however, more severe in the earlier than the later 1860s. It should also be noted that mortality from *typhoid* remained high. |

London than in Paris, Brussels, Berlin and Stockholm. Among major European cities only Vienna had a better record.[17] Similarly, in the 1890s, when London was compared with cities such as Leeds, Nottingham, Liverpool, Manchester and Salford, several of which were already obtaining their water from sources believed to be considerably less contaminated than either the Thames or the Lea, the capital appeared to be relatively exempt. As table 6.3 reveals, only Bristol, Birmingham and Bradford, among the larger urban centres, had lower rates.

**Table 6.3** Annual mortality from typhoid per million population in selected urban areas. Source: *RC Water Supply Metropolis* PP 1893–4:XL(II):Appendix G.I. Table II.

| 1871–80 | | 1881–90 | |
|---|---|---|---|
| London | 241 | Bristol | 145 |
| Liverpool | 297 | Birmingham | 177 |
| Bristol | 327 | Bradford | 183 |
| Birmingham | 330 | London | 187 |
| Newcastle | 357 | Sheffield | 196 |
| Manchester | 358 | Leicester | 209 |
| Leicester | 373 | Newcastle | 213 |
| Bradford | 410 | Hull | 223 |
| Sheffield | 435 | Liverpool | 236 |
| Salford | 438 | Manchester | 251 |
| Leeds | 456 | Leeds | 278 |
| Hull | 469 | Nottingham | 341 |
| Nottingham | 469 | Salford | 363 |

Yet the rate of decline, as distinct from the absolute level, of typhoid in London between 1870 and 1900 was disappointing when compared with other cities against which the performance of the capital could best be judged. The engineer to the East London Company who in 1900 claimed that West Ham had rooted out the infection and would therefore make a suitable location for a sanatorium reflected widespread late-nineteenth-century optimism.[18] But the sombre assessment of the former medical officer of St James, Westminster, Edwin Ray Lankester—'Typhoid we have always with us'—gave a more accurate impression of the continuing endemic significance of the disease.[19]

When we examine district death rates from the infection in London between 1870 and 1910, attention needs to be given to individuals normally resident in one part of London who were now dying in increasing numbers in an isolation or other hospital quite distant from their 'home' district. Adjustments of this type have been made to the data presented in table 6.4.[20] Although there are apparent anomalies—the surprisingly high mortality in Greenwich throughout the period and the poor performance of Westminster during the 1880s—the dominant pattern is a predictable one. The core districts of the East

**Table 6.4** Typhoid mortality per 100 000 population in the London Registration Districts, 1871–1900. Source: *Annual Reports* of the Registrar-General.

|  | 1871–80 | 1881–90 | 1891–1900 |
|---|---|---|---|
| Kensington | 21 | 22 | 10 |
| Chelsea | 29 | 16 | 13 |
| Westminster† | 32 | 23 | 27 |
| Marylebone | 27 | 18 | 13 |
| Hampstead | 20 | 19 | 11 |
| St Pancras | 29 | 23 | 16 |
| Islington | 32 | 24 | 13 |
| Hackney | 33 | 27 | 13 |
| St Giles | 23 | 19 | 13 |
| Holborn | 30 | 19 | 16 |
| City | 48 | 29 | 18 |
| Shoreditch | 34 | 20 | 16 |
| Bethnal Green | 36 | 21 | 19 |
| Whitechapel | 42 | 39 | 13 |
| St George-in-the-East | 35 | 17 | 17 |
| Stepney and Mile End | 32 | 25 | 18 |
| Poplar | 32 | 24 | 19 |
| St Saviour, Southwark | 32 | 15 | 17 |
| St Olave, Southwark | 34 | 20 | 17 |
| Lambeth | 35 | 25 | 12 |
| Wandsworth | 27 | 16 | 12 |
| Camberwell | 32 | 15 | 10 |
| Greenwich | 41 | 30 | 20 |
| Lewisham | 25 | 12 | 8 |
| Woolwich | 34 | 12 | 4 |
| London | 30 | 18 | 12 |

† This is a 'composite' district and includes St George, Hanover Square, Strand and Westminster. It makes allowance for numerous boundary changes.

End and the inner city, characterised by extreme poverty, over-crowding, continuing lack of access to constant supply, and severe environmental deprivation, experienced the highest levels of mortality. In some of these areas, such as Shoreditch, Bethnal Green and St George-in-the-East, improvement could be noted during the 1880s; in others, however—and here Whitechapel is a clear-cut example— typhoid continued to be a serious problem throughout the period 1870–90. It was, in fact, only during the first decade of the twentieth century that differentials between those districts least and most severely afflicted by the infection began to be eroded.

That it was necessary to attempt to exclude one means of spread while investigating every other possibility had been widely recognised by public health workers in the 1870s. By the end of that decade a majority of medical officers had become convinced that the storage and filtration plant which had been installed by the companies since the middle of the century had proved itself relatively effective.[21] Accordingly, other modes of infection—by milk, person-to-person contact, insanitary domestic arrangements or via sewage-impregnated shellfish—were scrutinised.

According to twentieth-century epidemiology, the role of infected milk in typhoid, as well as of milk which has been unsafely diluted with previously contaminated water, is well established.[22] Both these modes of infection, which are frequently deeply intermingled in an outbreak, were described in London in the second half of the nineteenth century. In 1870 Edward Ballard completed his classic survey of a North London epidemic which was traced back, in part, to a contaminated milk supply. 'We take a great deal of trouble', he wrote, 'to secure the purity of the water in dwelling houses, and to guard against its contamination from house-drain emanations, and from the emanations from cesspools; but with all our care a wholesale poisoning may take place because the article received into the houses and used as milk is diluted with water mixed with the contagion of typhoid fever.'[23] A similar process underlay an even more thoroughly documented epidemic in 1874 when, within the space of nine weeks ending on 30 August, 244 cases including 26 deaths, many of them in well-to-do households, occurred in Marylebone, St George, Hanover Square, Paddington, Hampstead, St Pancras, Soho and Kensington.[24] The Medical Department inspectors Netten Radcli.fe and W H Power discounted direct, water-transmitted infection, on the grounds of the 'extreme improbability of concurrent and adjacent local pollutions in the districts of the two companies [the West Middlesex and Grand Junction]', and proceeded to locate the origins of the outbreak at a dairy which was served by a wholly inadequate and insanitary water supply.[25] 'Until', they concluded, 'this habitude of using, and indifference to the use of, foul water or of water liable to be fouled, for dairy purposes is put check to, probably the most important source of infection of milk with enteric fever material will hardly be removed.'[26] Commonsense attention to cleanliness at dairies was insufficient: it was also imperative to legislate for a system of control administered by sanitary authorities both at farms and inner-city corner shops and grocers which sold milk.[27]

But legislation was delayed and metropolitan milk-transmitted typhoid continued to take its toll. Thus polluted and/or improperly over-diluted milk was indicted at Hampstead in 1877.[28] In 1879, however, James Edmunds, medical officer for St James, Westminster, who had recently located a milk-spread epidemic, was still emphasising the diagnostic difficulties which hampered this kind of operation. 'No chemical or microscopical examination', he complained, 'suffices to differentiate milk which has been made poisonous by such pollution.'[29] In 1883 several retailers in the Camden and Regent's Park subdistricts were believed to have been responsible for an extremely serious outbreak which claimed 132 lives—it was later found that each of the dealers had connections with the same farmer in Hertfordshire.[30] In 1891, again in Hampstead, there were 72 cases which could be directly related back to a substandard milk supply.[31] It is not possible to assess what proportion of aggregate mortality from typhoid for the period under review should be attributed to various categories of milk infection, but there is little doubt that large numbers of cases and fatalities were involved: and it was only in the 1890s, as milk quality slowly improved, and as public analysts began to work in closer collaboration with district sanitary authorities, that such outbreaks waned in frequency and virulence.[32]

It is equally difficult to give exact quantitative expression to the role of infected shellfish. This medium was first suspected and publicised by Arthur Newsholme, who began his career as medical officer at Brighton in 1888, and who was thus well placed to observe the filthy conditions in which offshore mussels, oysters and whelks either flourished naturally or were cultured artificially. Newsholme believed that shellfish contamination, as distinct from a continuingly unsafe water supply, had been a crucial factor in the failure of typhoid rates to decline more rapidly than they actually did during the final 15 years of the nineteenth century. 'During the whole period 1875–1910', he recalled, 'both Brighton and London had water supplies which can be absolved from the suspicion that they contributed to typhoid infection, and during the same period sanitation was steadily improving.'[33] Such a position in relation to water pollution was undoubtedly over-optimistic, but in nearly every investigation of local typhoid outbreaks in London in the 1890s shellfish were identified either as an unequivocal or as a highly likely contributory causal medium.[34]

Given the difficulties of proving the epidemiological course of milk- and shellfish-induced infection, it is hardly surprising that, in aggregate terms, a majority of deaths in typhoid outbreaks during the second half of the century were ascribed to insanitary domestic arrangements. A frequent complaint was that water which had been satisfactorily treated by the companies was subjected to domestic pollution in unattended and uncovered butts or in cisterns which were only rarely cleaned.[35] Families who were too poor, or who moved house too frequently, to take advantage of constant supply, were, as we have seen, accused of undoing the work of subsidence and filtration in which the companies had been investing since the mid-1850s. The response to

such charges of personal irresponsibility was that, through their powerful voice in Parliament the companies were in a position to modify
legislation in such a way as to make it exceptionally difficult for the
poorer areas to be assured of constant supply.[36] In this sense, so the
argument ran, the purveyors of water were indirectly responsible for
much water-transmitted disease. By agreeing to 'adequately and
effectually' filter their water under the Act of 1852, and to submit to
slightly stronger, although still almost wholly nugatory, government
inspection in 1871, they had contributed to the reduction of 'filth'
diseases. But, by refusing to liberalise the conditions under which
constant supply would be granted, they had partially nullified their
own efforts. The statistics were indisputable: as late as 1892 only 69 per
cent of all houses in 'water London' were in a position to dispense with
internal cisterns for drinking water.[37]

Even in the houses of the relatively well-to-do, the proximity of pipes
supplying water for sanitation and for cooking and drinking made
person-to-person infection within the individual dwelling a constant
threat. Randomly scattered outbreaks of this type, many of them affecting the professional classes, received frequent and puzzled attention.[38]
Such epidemics depended until a remarkably late date for their explanation upon one or another version of the 'vapour theory' of typhoid
transmission. (A distinction should be made between these attacks and
those which were being increasingly perceived in terms of specificity,
and the germ theory of disease.[39]) But where 'soil water' was observed
to drop into the cistern which contained a family's drinking water;
where a sanitary inspector noted the 'in-sucking of excrementitious
matter into the service pipe'[40]; and where house drainage systems for
sewage and storm water were inadequately separated[41] the role of
deficient plumbing and building in spreading typhoid required no
further emphasis. 'In the majority of cases', wrote the medical officer for
Kensington in 1873, 'the overflow of water in the case of accident to the
supply apparatus is prevented by means of a water pipe, which in many
cases is in direct communication with the house drains, without the
intervention of any trap. The waste pipe then serves as a ventilation,
often the only one, to the drains and the common sewer; and the use of
water, impregnated with sewer gases, may occasion diarrhoea, and, in
special circumstances, Typhoid Fever.'[42]

By the 1880s it had become clear to some of the more experienced
medical officers that typhoid in London had not been fully explained by
the local studies which had been undertaken—albeit with considerable
expertise—since the 1860s. It had proven exceptionally difficult to
produce convincing environmental diagnoses of more than a small
percentage of cases in a majority of local epidemic, or subepidemic,
outbreaks; and too often the district epidemiologist resorted to an

indictment of generally insanitary domestic conditions, even when it could be shown, following an order by a sanitary inspector and the rectification of an alleged deficiency, that fever frequently returned a year or so later to the same house, street or district.[43] The disease had been endemic and sporadically epidemic in Shoreditch since detailed records had first been kept in the mid-1850s, but identification of environmental transmission had been rare.[44] In Kensington in 1883, nineteen houses in twelve streets were attacked and there were five deaths, but the underlying pattern remained obscure.[45] In the Strand in 1895 a typical small-scale attack was broken down into the following categories: 'residual' from a previous outbreak; contraction of the disease outside the area of immediate epidemiological inquiry; the influence of defective drainage; and the consumption of infected shell-fish.[46] In the absence of more systematic interdistrict cooperation to trace infected 'migrants', and more reliable criteria for unequivocally defining 'defective drainage', such statistical analyses were no more than approximate and might actually mask the operation of a single and temporarily dominant means of spread.

By the early 1890s district data were again being analysed in detail by public health officials who were questioning the optimism both of the water companies and of an influential section of the scientific establishment. The persuasiveness of this more pessimistic attitude was, however, vitiated by the continuing unreliability of bacteriological techniques and the fact that no investigator had yet isolated the causative bacillus either during or immediately after a water-transmitted outbreak. Although by this time there was a growing commitment to the idea that specific diseases were frequently spread in polluted water, methods for identifying pathogenic micro-organisms were still underdeveloped. A precondition for more effective epidemiological fieldwork was the creation of public health bureaucracies to undertake routine, day-to-day observation. But methods for confirming connections between unsafe water and the incidence of typhoid were, as they still are, elusive. 'The incidence of typhoid fever', a mid-twentieth-century waterworks manual states, 'constitutes probably the most significant and ultimately accurate measure of the sanitary quality of a community water supply.'[47] But the importance to be attached to an improved supply as an explanation of reduced mortality from the infection needs to be weighed against other, contemporaneous, modifications to the environment and juxtaposed against the experience of neighbouring communities where the quality of water has remained unchanged.[48] As a pioneering epidemiologist noted, 'the failure of the method of overt correlation in space and time [is not] conclusive evidence that water was wholly innocent, any more than the presence of such correlation is evidence that water is guilty'.[49]

A number of influential public health specialists, including Shirley Forster Murphy, Edward Klein, Sims Woodhead and Edward Frankland, attempted to reverse the orthodoxy that continuingly and disturbingly high levels of typhoid were unrelated to failings in water purification processes. Although he believed that unsafe water may have been the major determinant of the infection over a much longer period than was usually recognised, Murphy was convinced that polluted supplies had been at least partially responsible for its recent prevalence.[50] Unwilling to accept conveniently monocausal explanations, he insisted that water, milk, flies and person-to-person infection were all likely to be implicated in wide-ranging outbreaks. He also raised the possibility—a novel one at the time—that epidemic fluctuations might be partly explicable in terms of shifts in the virulence of the typhoid bacillus itself.[51] Murphy attempted to substantiate his position by means of a statistical analysis of comparative metropolitan typhoid rates according to district and source of water supply. He was willing to admit that the 'London death rate from enteric fever bears favourable comparison to this rate in other towns having public water supplies which are not excrementally polluted'.[52] But what disturbed him was that districts supplied by companies drawing their water from the Lea were perceptibly more vulnerable than those served by companies drawing from the Thames: 'the tendency of this disease to attack more heavily the Lea-supplied population is only in part capable of explanation by such information as can be obtained as to the greater poverty of the population'.[53] Because many registration districts were supplied by more than one company, it was necessary to divide 'water London' into seven areas—'Kent', 'Thames', 'Kent and Thames', 'New River and Thames', 'New River', 'New River and East London', and 'East London'. Murphy then compared annual death rates from typhoid for each water district with an index of 'comparative mortality' for the whole of London. This procedure is set out in table 6.5. As a final test, Murphy compared the incidence of a range of infectious diseases, including typhoid, in 'Lea' and 'non-Lea' areas, making allowance for the possible effects of differentials in levels of poverty[54] and concluded that the incidence of typhoid correlated less significantly with poverty than with source of water supply (see table 6.6). A relatively rich man, in other words, drinking water derived from the Lea, was more likely to contract typhoid than was a relatively poor man drinking water from the Thames. And, of the diseases thus analysed, such a connection was observable in the case of typhoid, and typhoid alone.

Those who dissented from this interpretation argued that Murphy's statistics were fallible and that the London County Council, which was deeply embroiled in a continuing struggle over the control of the capital's water supply, had a vested interest in publicising typhoid and

**Table 6.5** Water supply and mortality from typhoid in London in the 1890s. Source: *RC Water Supply Metropolis* PP 1893–4:XL(II):Appendix C17:Table IV:222.

| Water companies | Mean population 1885–92 | Typhoid death rate (per million) | Comparative mortality (London = 100) |
|---|---|---|---|
| Kent | 122 533 | 89 | 65 |
| Thames | 1 683 617 | 119 | 88 |
| New River and Thames | 366 355 | 137 | 101 |
| Kent and Thames | 487 287 | 135 | 99 |
| New River | 496 909 | 147 | 108 |
| New River and East London | 458 723 | 155 | 114 |
| East London | 504 063 | 179 | 132 |

discrediting the companies.[55] But every available index revealed that a degree of pessimism was justified: the 1890s, as figure 6.3 reveals, were characterised by disquietingly high death rates from the infection. In 1894 there was an unusual prevalence of the disease between the forty-ninth and fifty-first weeks of the year—precisely the period when the 'autumn fever', dampened by cold weather, normally declined.[56] Once again Murphy sought and then rejected a variety of explanations—that the reversal of trend was a national rather than an exclusively metropolitan phenomenon, and that a wide range of infective media may have been involved—before fixing on an explanation which implicated unsafe water. He concluded that the widespread nature of the epidemic (only the East End and those supplied by the Kent Company had remained substantially unaffected[57]) pointed to reliance on inferior water which had been allowed into subsidence reservoirs during a period of flood and abnormally high demand. In the same year, and throughout the decade, sewage-contaminated oysters were also widely suspected but it had now become clear that this mode of transmission posed formidable epidemiological difficulties—it was, for example, by no means evident that shellfish consumption was more widespread among typhoid sufferers than among the population at large.[58] There was, furthermore, opposing evidence which suggested that abnormally severe river pollution, and weaknesses in subsidence and filtration, were probably more culpable than food poisoning. 'This [outbreak]', commented the medical officer of St George, Hanover Square in 1895, 'follows upon the delivery of insufficiently filtered Thames Water when the river is in flood, which was especially noticeable last year, the samples taken in November being very bad indeed. Some cases of the disease in London and elsewhere were believed to

**Table 6.6** Water supply, poverty and mortality from typhoid in London in the 1890s. Source: *RC Water Supply Metropolis* PP 1893–4:XL(II):Appendix C17:Table IV:222.

| | 19 districts receiving water from Lea | | 20 districts not receiving water from Lea | |
| --- | --- | --- | --- | --- |
| | Comparative mortality of 12 districts *above* average London poverty | Comparative mortality of 7 districts *below* average London poverty | Comparative mortality of 7 districts *above* average London poverty | Comparative mortality of 13 districts *below* average London poverty |
| Typhoid | 113 | 116 | 103 | 85 |
| Diarrhoea | 115 | 89 | 109 | 92 |
| Scarlet fever | 122 | 96 | 129 | 89 |
| Whooping cough | 114 | 93 | 116 | 90 |
| Measles | 121 | 91 | 115 | 87 |
| Diphtheria | 114 | 103 | 87 | 94 |
| All causes (all ages) | 110 | 97 | 110 | 92 |
| Infant mortality | 105 | 97 | 105 | 96 |

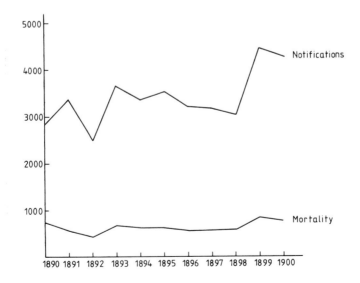

**Figure 6.3** Typhoid notifications and mortality, London 1890–1900. Source: *Annual Reports* of the Medical Officer of Health to the London County Council.

have been traced to the consumption of sewage polluted oysters, but this cause would apply equally to September and October.'[59] The floods of 1894 were followed by drought in the summer of 1895 and the companies were once again widely suspected of devoting insufficient attention to the delivery of wholesome water. During the 'water famine' of that year the East London Company was forced to admit that it had purchased unfiltered deep well water during a period of excessive demand, thus reviving memories of its grosser irresponsibilities in 1866.[60] Filtration techniques were now questioned on every side. 'Reliance', commented the medical officer for the Strand in 1896, 'cannot be placed upon sand-filters to protect communities from water-borne disease. . . . So long as we draw water from polluted sources like the Thames, we must expect at times outbreaks of various intestinal disorders, and should the water at these particular periods happen to contain the germs of cholera or enteric fever, epidemics of these diseases will certainly ensue.'[61] At their best, Thames and Lea water were admitted to be safe: at their 'flood-time worst' they were 'dubious'.[62] Too little warning was said to be provided by still clumsy and time-consuming chemical and bacteriological tests and, as we have already seen, London was now thought to be vulnerable to the kind of virulent typhoid which had struck Maidstone in 1897, an epidemic which 'set every Authority a question to answer. What security have they that the means adopted to protect their district from such an outbreak are sufficient?'[63]

By the end of the century the London County Council, the water companies, public health officials and scientists were in frequent dispute about the susceptibility of the capital to typhoid. One such conflict, between a medical officer and the water companies in 1899, when typhoid notifications for the capital as a whole rose to nearly 4500, reveals the administrative and scientific issues which now surrounded the question of water quality and disease prevention in London. In December 1898 George Corfield, the medical officer of health for St George, Hanover Square, wrote to the Chelsea and Grand Junction companies, complaining of the 'decidedly inferior quality' of water which had been distributed during the year. The companies replied, dismissively, not to Corfield but to the vestry committee to which he was formally responsible. Corfield then widened and deepened the area of dispute, alleging that parishioners had been complaining to him regularly, especially in the winter months, for over twenty five years. The companies were sufficiently embarrassed to make an official reply through their chemists, William Crookes and James Dewar. The argument now continued at a more overtly scientific level, with the company chemists denying Corfield's supposition that there was a measurable relationship between the 'organic' pollution of river water and typhoid. It was, they insisted, the microbiological quality rather than the quantity of 'polluting matter' which determined whether or not the disease would be transmitted. Corfield was willing enough to accept the changed premises which by then underlay the dispute and went on to cite reports by Edward Frankland and the Joint Public Health and Water Committees of the County Council that, even after filtration, company water frequently showed more than the recommended bacteriological maximum of 100 microbes per cubic centimetre. He added that two other independent chemists had reported adversely on the microbiological quality of water delivered by the Chelsea Company from November 1897 to November 1898 and had also cast serious aspersions on the effectiveness of its filter beds. He complained that Crookes and Dewar had introduced a diversionary tactic by implying that during the investigation of the notorious Maidstone epidemic Sims Woodhead and Washbourne had endorsed the view that, since no typhoid bacillus had been isolated, the water supply might therefore be declared innocent. Precisely the opposite, Corfield retorted: what the bacteriologists had in fact concluded was that the failure to find a specific organism in the implicated water was 'in no way to exonerate it'. The companies, as was their right, declined to answer Corfield's accusations and the correspondence then lapsed: unless the Local Government Board chose to institute an official enquiry the companies were under no obligation to deal with day-to-day complaints either from individuals or public health officials.[64] The conclusion which may be derived from this interchange is a familiar one—the medico-scientific establishment in London at the end of the nineteenth century was divided between those who believed that, in epidemiological terms, the worst was long past and would never return and an altogether more sceptical group which championed ever more stringent controls over the treatment and delivery of domestic supplies. This is well illustrated by the positions of William Ogle and Sims Woodhead in 1893. Ogle was ready to trust to what he called the 'light of experience'. London, he contended, had a remarkably satisfactory typhoid record. Up-river outbreaks had not been communicated downwards to the capital; even if dangerous bacilli found their way to the company intakes, it was unlikely that any individual would swallow enough of them to contract the disease; and the Thames and the Lea were undergoing a slow but certain process of purification.[65] Although modest about the achievements of bacteriology, Woodhead warned of the dangers of reliance upon the 'light of experience' and of employing statistical data from the past to produce predictions about the incidence of disease in the future. 'Although', he said, 'the conditions may be such that you will have no outbreak of typhoid fever, . . . , for a very long time, still the water may be in such a condition, and may contain such materials, that if typhoid bacillus happened to find its way into the water that water might become exceedingly dangerous.'[66]

These debates about typhoid and its connections with unsafe water became less intense as the disease declined to much lower levels after about 1905. The final reduction is best explained neither in terms of municipalisation nor the disinfection of public water supplies in the capital[67]—and chronological support for the latter hypothesis is weak since chlorination was only systematically introduced in 1915.[68] It is more likely that rapid hospitalisation of localised cases and more tightly controlled filtration rates following the troubles of the 1890s combined first to stabilise and then to force down mortality and morbidity. By 1904, Shirley Forster Murphy, reversing the methodology he had advocated a decade earlier, was turning his attention away from the impact of unsafe water *per se* and back towards poverty. 'The decline in the case-rate, death-rate and fatality of enteric fever', he now argued, 'has been maintained but some London districts suffer practically twice as much as others. This may in part be due to difference in food supply, but in all probability it is especially dependent upon other conditions of life and especially the habits of the population which give opportunity for the disease to spread from one person to another.'[69]

From about 1870 onwards many of those professionally concerned with the prevention of disease in London looked upon typhoid as an indication that a given water supply was unfit for human consumption. But this conclusion masks both the complexity of the behaviour of the disease itself and the ambivalence of a body of medico-environmental theory which had not yet made an unambiguous break with 'vapour' and miasmatic explanations of the dominant mode of transmission. Whenever company supplies seemed to have attained a degree of purity which would have eliminated the infection in its water-transmitted form, typhoid would unerringly recur in a series of small-scale but by no means trivial outbreaks. The full weight of investigatory expertise would then be brought to bear on epidemics which might kill as many as fifteen people and make four or five times that number very seriously ill. A convincing explanation—substandard milk, tainted ice-cream, unsafe water-cress, polluted shellfish or rotten food—might be found, but in the many instances in which the bacillus outwitted professional knowledge there was a tendency to resort to non-specific accounts based on the activity of 'noxious vapours'.

The most convincing analyses of late-nineteenth- and early-twentieth-century outbreaks of the disease pointed unequivocally to multicausal processes, and in the case of the outstanding researches of

Shirley Forster Murphy, to a delicate balance between intermittently unsafe water and life styles associated with and determined by poverty. Less painstaking field workers too easily assumed that since water was substantially safer than it had been 30 years earlier, small-scale epidemics ought never to be attributed to temporary, localised short-comings in company supplies. But late-twentieth-century epidemiological theory strongly suggests that many of these 'inexplicable' episodes were in fact probably primarily attributable to short-term deficiencies in water purification procedures. The perspective of many late-Victorian investigators was narrowed by the assumption, reinforced by the fate of the Prince Consort in 1861, that of all the 'fevers' of the second half of the nineteenth century it was typhoid which posed the gravest threat to the health and stability of the upper and middle classes: and evidence presented in this chapter does indeed confirm that the suburbs were regularly afflicted by the disease. But in the aggregate it was the poorest districts and the poorest inhabitants of London, as of every other urban region, who suffered most traumatically from the 'autumn fever'.

In the longer term, the regularity and frequency of typhoid epidemics; the unpredictability and complexity of the course of any given outbreak; and the multicausal mode of transmission were decisive in shaping the theory and practice of the profession of epidemiology.

1   This is an under-researched topic but see Charles Creighton, *A History of Epidemics in Britain* (revised edition, 1965) Vol 1, 70–1 and 165; and J F D Shrewsbury, *A History of Bubonic Plague in the British Isles* (Cambridge, 1971) 173 and 381.
2   F S Stewart, *Bigger's Handbook of Bacteriology* (eighth edition, 1962) 328 *passim*.
3   R Thorne Thorne, *The Progress of Preventive Medicine during the Victorian Era* (1888) 26; and E W Goodall, *A Short History of the Epidemic Infectious Diseases* (1934) 86.
4   See Budd's comments to the *Royal Sanitary Commission* (1871) Q 9235.
5   Cited in Adam Patrick, *The Enteric Fevers* (Edinburgh, 1955) 24.
6   *Report of the MOH: Hackney* (1878) 4; and Arthur Newsholme, *Fifty Years in Public Health* (1935) 347.
7   William Bulloch, *The History of Bacteriology* (Oxford, 1938) 266–7.
8   Creighton (note 1 above) Vol II, 160 *passim*.
9   J N Radcliffe, 'Reports on Epidemics', *Trans. Epid. Soc.* **ii** (1863) 411; *Report of the MOH: Shoreditch* (1862) 19; *Report of the MOH: Strand* (1859) 11; and *Report of the MOH: Holborn* (1866) 53–4.
10  This is discussed at greater length in my paper in Woods and Woodward (eds), *Urban Disease and Mortality in Nineteenth Century England* (1984) 105–10.

11    A summary of the underlying issues at stake is provided by A J Taylor (ed) *The Standard of Living in Britain in the Industrial Revolution* (1975).

12    It is likely that deaths from what the Registrar-General termed 'simple continued fever' should be included in mortality from typhoid: if this addition were made, it would involve a slight upward adjustment, particularly for the 1870s. On this point see G B Longstaff, 'The Seasonal Prevalence of Continued Fever in London', *Trans. Epid. Soc.* (1884–5) 72; and Charles Murchison, *A Treatise of the Continued Fevers of Great Britain*, third edition (edited by W Cayley, 1884) 682.

13    Thomas McKeown and R G Record, 'Reasons for the Decline of Mortality in England and Wales during the Nineteenth Century', *Population Studies* **16** (1962) 118.

14    George Rosen in H J Dyos and Michael Wolff (eds), *The Victorian City: Images and Reality* Vol II (1973) 635.

15    Major Greenwood, *Epidemics and Crowd Diseases* (1935) 157–8.

16    See, for example, G B Longstaff, 'The Recent Decline in the English Death Rate considered in connection with the Causes of Death', *J. Stat. Soc.* **xlvii** (1884) 227.

17    A Palmberg, *A Treatise on Public Health and its Applications* (translated by A Newsholme, 1893) 513–19.

18    *RC Water Supply Metropolis* (1900) Q 16 386. Evidence of W B Bryan.

19    *Ibid*, Q 10 636.

20    Table 6.4 has been compiled by reworking the Registrar-General's highly unsatisfactory material in the light of data contained in the Minutes of the Metropolitan Asylums Board and the Annual *Reports* of the Statistical Committee of that body. No adjustment has been made for the final decade under consideration since, by this period, absolute levels of cases had reached very low levels.

21    See, for example, *Report of the MOH: Marylebone* (1870) 23; and *Report of the MOH: Kensington* (1881) 123.

22    G P Gladstone, 'Pathogenicity and Virulence of Microorganisms' in Howard Florey (ed), *General Pathology* (third edition, 1962) 692–722; and John Ritchie, 'Enteric Fever', *Br. Med. J.* **2** (1937) 166.

23    *Report of the MOH: Islington* (1870) 11.

24    *Report of the MOH: St George, Hanover Square* (1874) 68–9; and 'Report on an Outbreak of Enteric Fever in Marylebone and the Adjoining Parts of London by J. Netten Radcliffe and W. H. Power' in *Supplementary Report of the Medical Officer of the Privy Council and the Local Government Board* PP 1874:XXXI:Appendix 6:137 *passim*.

25    *Ibid*, 144.

26    *Ibid*, 164.

27    *Ibid*, 165.

28    *Report of the MOH: Hampstead* (1877) 45–6.

29    *Report of the MOH: St James, Westminster* (1879) 32.

30    *Report of the MOH: St Pancras* (1883) 15–18; and *Report of the MOH: Islington* (1883) 54–6.

31    *Report of the MOH: Hampstead* (1891) 30–1.

32    For the effects of the various Milk and Dairies Orders issued by the Privy

Council under the Contagious Diseases (Animals) Acts see W M Frazer, *A History of English Public Health 1834–1939* (1950) 465; and on technological and marketing changes, E H Whetham, 'The London Milk Trade 1860–1900', *Econ. Hist. Rev.* **17** (1964–5) 369–80.

33  A Newsholme, *Fifty Years in Public Health* (1935) 205–6; and *idem*, 'The Spread of Enteric Fever by means of Sewage-Contaminated Shellfish', *J. San. Inst. GB* **xvii** (1896) 389–411.

34  See, for example, *Report of the MOH: Lambeth* (1894) 4–36 where watercress, as well as oysters, is implicated; *Report of the MOH: Fulham* (1899) 18–19; and *Report of the MOH: St Pancras* (1900) 39–41, for the effects of infected mussels.

35  See, for example, the graphic descriptions by Frank Bolton, 'Metropolitan Water Supply, 1871' PP 1875:XXXI:266.

36  R A Lewis, *Edwin Chadwick and the Public Health Movement 1832–1854* (1952) 329.

37  *Report of the Medical Officer of Health to the London County Council* (1892) 42.

38  Compare, for example, *Report of the MOH: Hackney* (1862) 13–14, and *Report of the MOH: St George, Hanover Square* (1890) 109–10 for continuing outbreaks in the 'better class' of house.

39  *Report of the MOH: St Pancras* (1891) 17–18.

40  *Report of the MOH: Marylebone* (1897) 13.

41  *Report of the MOH: Hackney* (1882) 6.

42  *Report of the MOH: Kensington* (1873) 69–70.

43  *Report of the MOH: Stoke Newington* (1897) 31.

44  *Report of the MOH: Shoreditch* (1877) 34.

45  *Report of the MOH: Kensington* (1883) 27.

46  *Report of the MOH: Strand* (1895) 126. For a comparative account of the difficulties of tracing the origins of typhoid to specific sources at a considerably later date see A K Chalmers, *The Health of Glasgow 1818–1925: An Outline* (Glasgow, 1930) 305–6.

47  American Waterworks Association, *Water Quality and Treatment*, second edition (New York, 1950) 37.

48  A Newsholme, *The Elements of Vital Statistics* (1923) 428–31.

49  M Greenwood, *Epidemics and Crowd Diseases* (1935) 159.

50  *Report of the MOH to the London County Council* (1893) Appendix 7, 5.

51  *Ibid*, 65.

52  *RC Water Supply Metropolis* (1893) Appendix C, Q 17 220.

53  *Ibid*, 227.

54  *Ibid*, 225. Murphy's definition of poverty was heavily influenced by the work of Charles Booth.

55  Arthur Shadwell, *The London Water Supply* (1899) 65–6.

56  *Report of the MOH to the London County Counci*(1894) Appendix II, Table I.

57  This was an intriguing matching since these were the two areas which had generated more controversy and anxiety than any others in London during the second half of the nineteenth century.

58  *Report of the MOH to the London County Council* (1902) 59–61; (1903) 33–5. For the difficulties of isolating shellfish contamination see *Report of the MOH to the London County Council* (1910) 42.

59    *Report of the MOH: St George, Hanover Square* (1895) 127.
60    *Report of the MOH: Hackney* (1895) 46–9.
61    *Report of the MOH: Strand* (1896) 139.
62    *Report of the MOH: Stoke Newington* (1897) 30.
63    *Ibid*, 31.
64    A summary of the correspondence upon which this account is based is contained in *Report of the MOH: St George, Hanover Square* (1899) 157–69.
65    *RC Water Supply Metropolis* (1893) Qs 10 300–10 316.
66    *Ibid*, Q 13 081.
67    It should, however, be noted that the typhoid bacillus is highly susceptible to chlorination. See H S Banks, *The Common Infectious Diseases* (1949) 326.
68    John C Thresh and John F Beale, *The Examination of Waters and Water Supplies,* third edition (1925) 106.
69    *Report of the MOH to the London County Council* (1904) 36.

# Part III

# A Degree of Control

(The law) was an arm of politics and politics was one of its arms

E P Thompson, *The Poverty of Theory and Other Essays* (1978), 96.

# 7 The Politics of Purification

In the northern industrial areas, the debate about river pollution which took place between the 1850s and the 1870s revealed deep-rooted differences between governing élites in the countryside and in towns. On the one hand, farmers and landed aristocrats strongly favoured the introduction of measures which would have compelled manufacturers to adopt less environmentally harmful production techniques and forced municipalities to invest in more effective methods of sewage treatment. National legislation of this type should be undertaken, they argued, before water meadows were transformed into foul-smelling marshes and livestock poisoned by refuse from lead and coal mines.

Manufacturers, Liberal parliamentarians and other apologists for *laissez-faire* capitalism took a quite different view. Characterising the landed aristocratic lobby as a narrow and indolent interest group, wholly concerned with aesthetic appreciation of the countryside and salmon fishing, they strenuously campaigned against legislation which would have raised costs of production and interfered, so they argued, with the 'creative collaboration' between capital and labour, employer and employee. John Bright was an eloquent advocate of this anti-aristocratic, anti-rural, classically liberal ideology. 'Was it not of the slightest interest', he demanded during the debate on the interventionist River Water Protection Bill of 1865, 'that 300 to 400 men and their families, who obtained a good living by their honest industry in a mine in the Welsh mountains were likely to be thrown out of work because small numbers of men who liked salmon fishing were seeking to place intolerable restraints on trade?'[1] Another Liberal, N Kendall, the member for East Cornwall, asked Lord Robert Montagu, the sponsor of the same bill, 'what when he spoke of the loss of fish as the loss of capital, did he imagine would be the loss to England of the closing of her mines'?[2]

In London, however, the confrontation over river pollution took a different and a more complex form. This was partly determined by the capital's dominant economic and social structures which remained obstinately non-industrial during the first great age of national industrialisation.[3] The absence of a unified Liberal manufacturing class, deriving its wealth from one of the great staple industries; the emphasis upon commerce and consumption, rather than production; and the early development of a thick, insulating belt of suburbia all softened the stark differences between urban and rural economic interests and values.[4] Another crucial difference was that in the North the rapid growth and social repercussions of industrialisation had conditioned reformers into concentrating upon the impact of manufacturing effluent. In London, on the other hand, the precocious growth of population from the late seventeenth century onwards had dictated that water pollution would be perceived predominantly in terms of human waste.[5]

For these and a number of other reasons, the history of river pollution and its control in London during the nineteenth century raises issues which are both more complicated and more elusive than those in the industrial North. In this chapter the haphazard development of agencies concerned with the management of the river following the traumatic crisis of the Thames in 1858 is viewed in the context of the troubled relationship between the national executive and the unreformed system of metropolitan government. Two bodies, the Metropolitan Board of Works and the Thames Conservancy Board, intense rivals, vying with one another for power and influence in a new sphere of municipal activity, are seen as representing divergent political values within that system. Both are also examined in relation to or reaction against the ingrained traditionalism of the City of London, at a time of vociferous political debate, both among Londoners and in Parliament, on how best to limit the power of the ancient corporation. In this sense, we see the Conservancy Board, like the City, with which it had close connections, priding itself—and here it is useful to draw upon the vocabulary of late-eighteenth-century political culture[6]—upon its 'independency' and resisting the notion that it should either represent or be made accountable to the great majority of Londoners. In its conflict with the Conservancy Board, the Metropolitan Board of Works represented, albeit in embryonic and at times pompously self-aggrandising form, that movement towards metropolitan democracy, representativeness and accountability which was to receive more coherent expression with the establishment of the London County Council in 1888. It is tempting to concentrate upon the Metropolitan Board's functional limitations, its lack of flair, and its final, ignominious juggling with rigged building contracts. But examination of environmental control in nineteenth-century London suggests that, in this field, the Board supported many of the administratively progressive principles which were later to be politicised and implemented by the London County Council.[7]

To an important extent, then, this chapter is concerned with conflicting ideals within the arena of nineteenth-century metropolitan politics. But we must not overlook the theme of regional contrast. In the industrial North, and in the west Midlands, the politics of pollution brought extra-urban governing élites into direct confrontation with the new industrial bourgeoisie in a manner which indirectly illuminated the larger struggle for state power.[8] In London, on the other hand, where, as we have seen, class formations were less monolithic, the battle was between vested traditionalism rooted in great commercial wealth—exemplified, institutionally, by the 'old corruption' of the City, and so also, its critics would claim, by the Conservancy Board—and newer forces of metropolitan populism and administrative efficiency. The struggle was a long one and it was closely and unyieldingly contested.

Following the dramatic environmental crisis of 1858, and the decision to construct a main drainage system for the capital, there was no single body possessing statutory powers to reduce pollution on the Thames. During the debate on the state of the river in 1858 it had been noted that there were four bodies which were 'armed with certain powers in the matter': the Board of Works, the Chief Commissioner of Works, the Thames Conservancy and the Sewage Commissioners.[9] The involvement of the Board of Works and the Chief Commissioner of Works was essentially short-term and directed towards enabling the Metropolitan Board to borrow adequate funds to finance the construction of the main drainage scheme. When this operation had been successfully carried through, the Metropolitan Board took over sole responsibility as the sewage authority. At very nearly the same time, another recently established body, the Thames Conservancy Board, which had been set up in 1857, was given jurisdiction over the Thames between Staines to the west and Yantlet Creek at the mouth of the river to the east; while a third, the older navigation authority, the Thames Commissioners, had its territorial responsibilities restricted to the upper river above Staines.

The price which the new Metropolitan Board had to pay for the assistance which it received from central government was a strict delimitation of its functions. Thus, although it was required by statute to construct the sewage system in such a way as to prevent further deterioration of the Thames it was not empowered to undertake complementary policing measures to control pollution. 'The plans for draining the metropolis', W Roupell, the Liberal member for Lambeth remarked in 1858, 'did not include any scheme for purifying the river,

for the Metropolitan Board had no jurisdiction over the Thames. It was
the duty of the Executive Government to deal with that question.'[10] But
the Government, which had already been attacked for what many
argued to be its complicity in the capital's failure to put its own house in
order, strongly resisted further involvement. The Metropolitan Board
had been granted access to loans large enough to enable it to construct
the main drainage system: the control of pollution, or what was sometimes optimistically referred to by contemporaries as 'repurification',
was not a task which either Derby's short-lived Tory ministry or
Palmerston's Liberal administration was willing to take on.

Like the Metropolitan Board of Works, the Conservancy Board had
been a child of compromise. In 1840 the Thames Commissioners,
primarily a navigation authority, had become involved in a tortuous
dispute with the Crown over the ownership of the foreshore of the river,
an argument sparked off by the then Whig government's determination
to embark on the construction of the Victoria Embankment. During the
eighteenth century the Commissioners had drawn their members
exclusively from those holding a property interest in the upper river
above Staines, but following an Act of 1770 the City of London had
gained a larger representation and influence, leading to a *de facto*
arrangement whereby a committee of extra-metropolitan Commissioners had superintended the navigation above Staines, and the City
had taken over responsibility for the river between Staines and the sea.
In effect, then, the dispute between the Commissioners and the Crown
was a dispute between the City and the Crown.[11]

It was only in 1856, following the expenditure of very large legal sums
on both sides, that a settlement was reached. According to this
compromise, the Crown waived all property rights relating to the
Thames; the Thames Commissioners retained their jurisdiction over
the upper river above Staines, and a new body, the Thames Conservancy, representing the City and the national executive, was entrusted
with the navigation and improvement of the river between Staines and
the sea.[12] It was hardly surprising that both the navigation and the
purity of the Thames had suffered very seriously during this sixteen-
year dispute and contemporaries were hopeful that the legislative
compromise of 1857 would halt yet further deterioration. 'It was
thought very desirable that the litigation should be terminated', the
Chancellor of the Exchequer, Sir George Cornewall Lewis, stated in
1857, 'inasmuch as the conservancy of the river was to a great extent
paralyzed from a want of those funds which were applied to the defence
of the corporation against the claim of the Crown.'[13] The Conservative
member for Oxfordshire, J W Henley, sought to calm the fears of those
who argued that the City would dominate the new board and dissipate
its slender financial resources on banqueting and ceremonial: the bill

establishing the Conservancy, he claimed, 'fully guarded against the corporation taking a farthing of the money for themselves, but every farthing would have to be laid out in the proper conservancy of the river'.[14] But sceptics were unconvinced and insisted that the Government's representatives on the Board would invariably be out-voted by vested interests.[15] 'Every step that had hitherto been taken', the Duke of Newcastle warned during the second reading of the Thames Conservancy Bill, 'had been in the wrong direction. The new body had been given into the hands of the Corporation of the City of London . . . and this had complicated the evils under which they were suffering.'[16] Opponents of the new board further contended that plans for halting environmental deterioration in the capital should be linked to a complete reform of metropolitan government as a whole. If it had been possible to put radical change of that type before the House, Sir James Graham insisted, 'the conservancy of the Thames would not be placed in such a body as was now suggested'.[17]

The membership of the Conservancy in 1857 consisted of the Lord Mayor, two aldermen and four Common Councilmen of the City of London, two nominees of Trinity House Lighthouse, two representatives of the Admirality, and one from the Board of Trade.[18] Following the Thames Navigation Act of 1866 eleven new members were added. Four of these represented the now finally defunct Thames Commissioners; two were spokesmen for substantial owners of shipping; two were nominated by the steam tug and lighter trade; and one new vote each went to the docking interest, the steam passenger trade, and the Board of Trade respectively. There could be no doubt that the City had been well able to maintain its position as the single most powerful influence over the new body; and it was for precisely that reason that the Metropolitan Board immediately launched a concerted attack against it.

In its final report in 1888 the Metropolitan Board would complain that 'it is a curious circumstance that the Board, as the governing authority in London, has not and never has had any control over the Thames, or any other voice in its management or conservancy. It seems as if every interest of a special and limited character was represented to the exclusion of the broad general interests of the inhabitants of London; for these can in no sense be said to be represented by the six Conservators nominated by the City Corporation.'[19] From a very early date the most persuasive proponent of the view that the Metropolitan Board should have had a seat on the Conservancy was the former body's long-serving chairman, Sir John Thwaites. Thwaites argued, firstly, that a modern river authority must represent and be seen to represent the interests of the community through which the river ran; secondly, that formal cooperation between the two boards would reduce the likelihood of conflict over the operation of the main drainage system; and thirdly, that the Metropolitan Board had a *de facto* interest in the state of the river as a whole.[20] Thwaites also cast doubt on the Conservancy's disinterestedness. 'The Conservancy', he argued in 1866, 'does not represent the metropolis, but it represents the steamboat companies, it represents the wharfingers, it represents the corporation of London, but it does not represent the metropolis.'[21] He alleged that the boating trade was frequently guilty of polluting the river and that therefore, on those occasions, the Conservancy was inevitably expected to take action against its own membership. 'If there is a liabilty to a nuisance', he contended, 'the parties most likely to commit the nuisance are not the parties who ought to form part of the Board whose duty it is to control and suppress the nuisance.'[22] The Conservancy remained unimpressed. To the charge that commercial and City-dominated factions were over-represented, it replied that a more 'democratic' constitution was unsuited to the needs of the river, adding in high, rhetorical, mid-Victorian style that 'Imperial interests of much importance are concerned in the management of the Thames, for which it is convenient to provide.'[23] The secretary of the Conservancy, Captain E Burstal, took particular pride in the fact that he only rarely corresponded with the Metropolitan Board. The vice-chairman, Sir Frederick Nicholson, went even further. 'They [the Metropolitan Board] have nothing whatsoever to do with the river, that I am aware of, except bringing sewage into the river at Barking: they keep it out of the river in a special district but they bring it into another district.'[24]

This aggressively legalistic statement underscored the Conservancy's view of the Metropolitan Board as no more than an officially authorised polluter of the eastern section of the river—an interpretation which was conveniently confirmed for them when, in 1869, the inhabitants of Barking, the suburb in which one of the outfalls of the main drainage system was situated, claimed that the health of their locality was being threatened by the proximity of large volumes of inadequately treated sewage.[25] The government inspector, the ubiquitous Robert Rawlinson, concluded that the complainants had exaggerated the nuisance and that the unsewered and generally unhygienic state of Barking itself was as great a potential danger to health as the main drainage outfall. In the interim, however, the Conservancy had taken up the towns-people's case and brought forward a bill which would have required the Metropolitan Board to deodorise all sewage before it was allowed to enter the river. The Metropolitan Board's response was that deodor-isation on such a large scale would have been unacceptably costly. A deadlock ensued and the affair was referred to a select committee. The committee attempted to reduce the tension between the two bodies by proposing a compromise which would have required the Metro-politan Board to finance any dredging costs which might be incurred by the creation of mud banks at the main drainage outfalls. By way of compensation, the Metropolitan Board was finally to be granted representation on the Conservancy; and were further conflicts to arise, referees, appointed by the Board of Trade, were to be empowered to mediate between the two bodies.[26] Now it was the Metropolitan Board which moved on to the offensive, claiming that since the quality of the water at the outfalls had undergone what it described as a 'spontaneous' improvement, it was no longer reasonable for the Conservancy to demand that the Metropolitan Board be required to undertake the expensive task of dredging the river.[27] And here the matter rested for very nearly a decade. Already, however, a state of armed peace had come into existence between the two bodies and the seeds of open warfare, always embedded in a legislative arangement which reflected the continuing fragmentation of metropolitan government as a whole, were to bear bitter fruit in the early 1880s.

Battle was rejoined in 1878 and this new conflict was to sustain itself until the final, ignoble collapse of the Metropolitan Board in 1888. It ranged over numerous issues which, had it not been for the acrimony which now existed between the two boards, might have been settled through the good offices of the capital's now mature and experienced scientific community. But by now mediation was exceptionally difficult to achieve. In the same year the Conservancy repeated its charge that dangerous mud banks, the by-product of the Metropolitan Board's allegedly anachronistic system of sewage treatment, were threatening the long-term quality of the river close to the outfalls at Crossness and Barking. The Conservancy's case was dramatised by the sinking in the lower river of a steamboat, the *Princess Alice*, with great loss of life, as well as by probably exaggerated reports that some of the passengers had quite literally choked to death on untreated sewage.[28] Since both parties to the dispute adopted characteristically stubborn attitudes, the Board of Trade appointed an umpire, who reported in 1880 that no case could be found against the Metropolitan Board.[29] Immediately, however, the grounds of dispute shifted yet again, with the Conservancy arguing in 1882 that the Metropolitan Board was still making use of a method of sewage disposal which was threatening the health of the inhabitants of Barking and Crossness. This was strenuously denied by the Metropolitan Board which claimed that the lower river had in fact recently undergone a substantial improvement.[30] The area of disagreement had now clearly become so general, and the possibility of a settlement so remote, that departmental mediation appeared unlikely to lead to reconciliation; and it was for this reason that a Royal Commission, with a brief to undertake a wide-ranging survey of sewage disposal for the entire metropolitan region, was appointed. The Metropolitan Board immediately claimed that two of the proposed commissioners had represented the Conservancy during an earlier phase of the dispute—

now, it seemed, even the appointment of an impartial investigatory body must be dogged by bickering.[31] Eventually, however, the Commission was able to set about its work, seeking evidence from a wide range of scientific and technical witnesses, making first-hand observations of its own by boat, and exploring the complex web of mutual recrimination which had characterised relations between the two bodies for more than 25 years. But before the Commission could present its report, the position of the Metropolitan Board was finally and fatally compromised when, during the unusually hot summer of 1884 supplies of deodorising chemical were shown to be quite incapable of stifling the stench of the river at the outfalls.[32] In the absence of an effective technique for the large-scale treatment of sewage, suburbanites were now subjected to an experience which was all too common in many urban communities during the nineteenth century: the trans-position of a pollution problem from central city areas, where it had finally come to be found intolerable by classes able to exert influence over public policy, to outer districts which had hitherto been relatively insulated from the impact of massive environmental deterioration.

The Commission then demanded that the Metropolitan Board introduce a more sophisticated method of treatment at the outfalls. The Board grudgingly complied, and in the period between the final report of the Royal Commission and the establishment of the London County Council as the sewage authority for the metropolitan region, it did indeed make use of a more efficient process based on large-scale precipitation, the regulated chemical treatment of effluent, and the final dumping of the remaining sludge far out at sea.[33] The demise of the Metropolitan Board in no way reduced the rancour which had now for so long attached to the issue of pollution control on the Thames. Indeed, during the 1890s the state of the river, like London's water supply, became more contentious and overtly politicised than at any other time since the 1850s. Confrontation was initially sparked by the progressive wing of the London County Council, but the Conservancy made no concessions and remained aloof from the new system of metropolitan administration: it continued to present itself as an independent and competent body which would in no circumstances demean itself by bowing to alien principles of democracy and account-ability. The County Council's primary aim was to gain the place on the Conservancy which had for so long eluded the Metropolitan Board. 'The Thames Conservancy Board', J Stuart, the progressive member for Shoreditch, told the House in 1893, 'consisted of 23 persons, and not one of them was in any sense directly or indirectly a representative of the 4,500,000 people who were so vitally interested in the Conservancy of the Thames.'[34] C B Stuart-Worsley, member for Sheffield, questioned, in tones reminiscent of the 1850s, the Conservancy's constitution and its expertise. 'The present constitution', he said, 'was not of an ideal character and seemed to have been conceived in the interest rather of the navigation than of water drinkers.'[35] John Benn insisted that the municipalities of the upper river were deeply hostile to the County Council and 'hoped he was not speaking disrespectfully of the persons who resided in the District of the Upper Thames when he said that they had attempted to use the river as a drain, and had compelled other people to drink the water'.[36] All this was distinctly more aggressive than the indictment which had been repeatedly directed at the Conservancy by the Metropolitan Board; and now, during the 1890s, the 'corruption' of the former body and the need for the 'democratic' control of anti-pollution agencies on the Thames were brought to the forefront of metropolitan political debate.

The Conservancy continued to be successful for a while in resisting progressive demands that it should in any sense 'represent' a metropolitan constituency. 'The Board would not be prepared', its Parliamentary Committee minuted in 1894, 'to adopt the principle of representation in proportion to population, as they consider that this principle is inapplicable to the Conservancy of the Thames, and that its adoption would lead to great injustice to important interests in and outside the Metropolis.'[37] But pressure for change was maintained and concessions were finally and inexorably won. As a result of the Thames Conservancy Act of 1894, the Conservancy was forced to give up four seats to the County Council. And yet, far from reducing tension between the two bodies, this change appeared to restoke the fires of conflict; and there were repeated and hostile confrontations in committee for the rest of the 1890s. Thus the Conservancy provocatively refused to grant the County Council members of the Board access to technical information relating to the collection of samples of water. (The representative of the Board of Trade was similarly rebuffed.)[38] The members of the Conservancy's River Purification Committee were so frequently at odds with the single County Council representative that the full Board felt compelled to issue a vote of confidence in the committee's competence and to publicise what it believed to be the inaccuracy of the attacks which the County Council had launched against it.[39] For a short period accusation and counter accusation ceased, only to flare up once more in 1897 in the aftermath of the Maidstone typhoid epidemic.[40] The County Council then began to consider the unified control of the capital's water supply, and the sources upon which that supply depended, as crucial to the protection of health throughout the metropolitan region. But how, the radicals argued, could a large-scale, water-transmitted epidemic be averted if anti-pollution measures continued to be carried out by a body which had shown itself to be the henchman both of City interests and élitist upper-river municipalities? Unmoved, the Conservancy Board continued to proclaim its 'anti-democratic' independence and its positive record in the field of pollution control.

Following the establishment of the Metropolitan Water Board in 1904, the 'water question' seemed momentarily to lose much of its animus—the expensive purchase of the private water company shares made the democratic control of utilities and the protection of public health less central to the progressive programme. And yet several major institutions remained immune to the new spirit of public surveillance. To the London progressive of the first decade of the new century the most blatantly delinquent of these bodies was undoubtedly the London Docks, the future of which could not be separated from that of the management of the river as a whole. The struggle for municipalisation proved to be a long and savage one, and the victory which the County Council finally secured in 1908 left many progressives deeply dissatisfied. Herbert Morrison, who dubbed the newly created Port of London Authority (PLA) 'a capitalist Soviet', lamented that the new body over-protected the interests of the dock-owners at the expense of both the County Council and the dockers.[41] Of central importance in the present context, however, is that a minor clause inserted into the Act establishing the PLA made that body, and not the Conservancy, responsible for all measures concerned with the reduction of pollution between Teddington Lock and the sea. Suddenly and peremptorily, the powers of the Conservancy were restricted to that part of the river which had once been the province of the eighteenth-century navigation authority, the Thames Commissioners. Moreover, the size of the reconstituted board, which had been raised to 38 in 1904, was reduced to 28, with the City now unequivocally controlling only two seats as against three directly nominated by the County Council and another three—two for the Metropolitan Water Board, and one for the PLA— over which the County Council clearly exercised a decisive influence. The County Council may have been partially outmanoeuvred by the dockowners; but it had finally and decisively turned the tables on the Conservancy. Belatedly and posthumously, the Metropolitan Board of Works had achieved a kind of revenge.

One further question needs to be asked; how successful was the Conservancy Board, derided and dismissed as it was by the Metropolitan Board of Works and latterly by the London County Council, in actually controlling pollution on the Thames between 1857 and 1908? There can be no doubt that in its earliest days the Conservancy lacked both the legislative means and adequate personnel to prosecute more than very small numbers of offenders. Forced to rely on the services of part-time detectives and spies, as well as on the costly help of the already over-worked Thames police force, its surveillance of the river between the late 1850s and the mid-1860s can have been no more than minimally effective.[42] Thus canny tradesmen paid bargees to dump waste into the Thames and organised mutual protection rings—one man keeping watch while others tipped rubbish quietly overboard.[43] Captains of steamers tacitly encouraged similar illegalities, in the knowledge that under the Act of 1857 it was the individual member of the crew, and not his superior officer, who was liable for prosecution[44] and that magistrates were known to be loath to impose fines on those least able to pay them.[45] It was against this background that, in 1863, the Board obtained a private act which secured it the right to exact heavy fines from offending barge owners[46] and also enabled it to employ two additional inspectors of nuisances.[47]

By the mid-1860s the Conservancy had come to perceive its role more in terms of reducing river pollution than improving the navigation; and this self-image was reflected in the debate on the Navigation Act of 1866 which, according to one involved contemporary, should have been renamed a 'purification act'.[48] The measure defined the overlapping rights and responsibilities of the Conservancy and the water companies—the companies were to make an annual contribution of £1000 each to the Conservancy for the right to draw water from their traditional sources in the Thames, while the Conservancy was to deploy these funds to reduce pollution.

From the mid-1860s onwards there was growing official anxiety about the scale of pollution on the upper river[49] and a developing realisation that, as the secretary of the Lea Conservancy forcefully put it, 'you cannot deal with these things unless you have a strong hand'.[50] From this period, in fact, it is possible to identify two complementary strands of thinking as to the future organisation of pollution control on the Thames: firstly, that the compromise of the late 1850s which had granted the Conservancy jurisdiction over the river from Staines to the sea, but which had deprived it of responsibility for that portion of the inner-city river over which the Metropolitan Board exercised its authority, was unsatisfactory; and secondly, that the water supply of the capital could only be protected if a river board were to be granted considerably larger powers of intervention—powers which contemporary advocates of the doctrine of local self-government would be bound to characterise as semi-dictatorial.[51] Many of those who supported a long-term solution, which, in the words of C D Acland, Liberal member for North Devon, would have brought 'the whole of the Thames under unity of management'[52] had been influenced by the recent and radical report of the Rivers Pollution Commissioners on the state of the Thames. The crucial recommendation here, and one which was likely to offend the multiplicity of individuals and bodies still claiming a material interest in the river, was that the Thames should 'be placed under the superintendence of one governing body'. The Commissioners argued that the officers of a board of this type should be allowed to make inspections without warning and to exact heavy fines for proven cases of pollution. They also insisted that local authorities should be given powers of compulsory purchase for the purpose of sewage treatment.[53]

In the event, as we have already seen, the Thames Navigation Act of 1866, together with a shorter measure in the following year, gave the enlarged Conservancy jurisdiction over the Thames from Cricklade in Wiltshire right down to the sea. The Board was now also empowered to inspect and take measures to reduce pollution on all watercourses within a three mile radius of the main river. But the question of the control of pollution on the inner-city Thames remained unsolved. The Metropolitan Board of Works was still under no obligation to take action against those who polluted the river within its area of juris-diction.[54] And this anomalous arrangement emphasised the admini-strative division of a river which, from the geographical and ecological points of view, was now increasingly coming to be seen as indivisible. It was also a system based on the barely tenable premise that, in an efficiently sewered urban region, additional anti-pollution measures were not required.

Nevertheless, throughout the 1870s and 1880s the Conservancy extended both its coverage and the additional personnel which this demanded. Scientific specialists—Edward Frankland, Henry Letheby, and William Odling—were employed as consulting analysts to report on waste matters which were believed to be pollutants.[55] Despite financial constraints, inspection of the upper river became more regular, and residual powers which had been vested in the Board to prevent municipalities on the upper Thames from passing untreated sewage into the river were brought into operation. Persuasion and technical advice were usually considered more effective than pro-secution: a vindication of this approach was that, by 1871, the recalcitrant and suspicious town of Reading had developed a system of sewage disposal which no longer seriously polluted the river.[56] The Conservancy now also encouraged authorities above Staines to begin to help themselves by means of affiliation to the Lower Thames Valley Main Sewerage Board, which had been set up in 1878, and by 1880 towns such as Oxford, Abingdon, Slough—and Reading—had adopted less environmentally harmful methods of disposal.[57]

And yet the management of the Thames, it was perceptively remarked in 1881, was still theoretically shared between no fewer than five bodies: surely it was now encumbent upon the Board of Trade, as a result of its statutory responsibility for water supply legislation in London, to cut the Gordian knot?[58] Other commentators made the point that it was anachronistic to continue to legislate for the river by means of local acts. As C T Ritchie, member for Tower Hamlets, pointed out, such measures had been designed to demarcate legislative functions within a specifically local context but 'the idea of calling the Thames through 200 miles of its course, touching as it does, many large and important towns, and conveying the trade and commerce of London upon it—the idea of calling it a 'locality' was simply ludicrous. One might as well call the Mississippi a locality.'[59]

The continuing inflexibility of metropolitan government dictated that advocates of a more integrated approach to the control of river pollution had to rest content with an extension of the Conservancy's scale of operation on the upper river. Under an act of 1878 it became empowered to inspect watercourses up to a maximum distance of ten miles from the main river—clearly a valuable extension to the 'three mile limit' of 1866. Fines for proven polluters were also increased and this necessitated the appointment of two more full-time inspectors, with rising administrative costs to be met partly out of increased tariffs paid by the water companies.[60] From the companies' point of view this was money well spent since the Conservancy was now giving attention to unsophisticated sewage disposal techniques in the area between Staines and Chiswick which, if left unattended, could have posed a serious threat to metropolitan supplies. From the mid-1880s onwards the inspectorate between Staines and Teddington was strengthened and the Conservancy took swift action against vested interests in the former locality who had been seeking to resist additional outlay on an improved sewage system.[61] By the end of the decade the administrative personnel on the river between Cricklade and Staines consisted of a full-time inspector, his assistant, and two river keepers; in addition to bicycles, this team now had access to a 'modern steam launch'.[62]

In 1894 the Conservancy finally gained control over the entire water-shed—a power which, as we have seen, it was to exercise until the traumatic year of 1908.[63] Since 1893 the river purification committee had been 'dealing with questions relating to the pollution of the river and its tributaries, and for prevention of such pollution, and to have control over the officers engaged in securing the purity of the Thames and its tributaries'.[64] The watershed was divided into seven districts with one inspector and one workman, employed on a structured salary scale including special allowances for travel, policing each of them. At the centre, a full-time analyst tested the samples which were collected in the districts.[65] The growing scientific and organisational maturity of the Board was reflected in the scale of routine intervention. 'During the past year', it was reported in 1895, 'all the towns, villages and smaller places have been visited by the inspectors, and the various authorities and persons responsible for pollution of the respective streams have been called upon to take measures for the prevention of such pollution.'[66] During 1898 'as many as 78 towns or villages, with an aggregate population of over 180,000' were said to have diverted polluting substances from streams which were tributaries of the Thames.[67] By the mid-1890s district metropolitan sanitary officers, who had hitherto conspicuously failed to make use of what limited powers they possessed to combat river pollution, were beginning to cooperate with the Conservancy's inspectorate and analyst.[68] And, by the end of the century, the analyst was testing an average of nearly six samples a day.[69]

Throughout the period between its establishment and its partial replacement by the PLA in 1908 the Conservancy Board consistently presented itself as an independent and scientifically progressive authority which was the servant of London, property-holders and municipalities on the upper river, and what it rhetorically referred to as 'larger Imperial interests'. The extent to which this self-image accorded with reality cannot be fully ascertained, although as later sections of this chapter have shown, there are grounds for believing that, by the 1880s, the Conservancy was protecting the river from much unnecessary pollution. For their part, the self-consciously reformist Metropolitan Board of Works, and latterly the London County Council, depicted the Conservancy as a corrupt and élitist body, constitutionally retrogressive, functionally inept, and always open to more or less overt manipulation by the City of London. 'Corruption' and 'efficiency' are historically relative terms and any attempt to balance out the veracity of the stylised rhetorics which the rival boards directed at one another is unlikely to succeed. A more fruitful approach, both to the emergence of pollution as a social problem and to institutionalised attempts to reduce its incidence is to examine values and polarities—the patrician versus the populist mentality, 'independency' versus accountability, administrative autonomy versus municipal centralisation—which coloured successive waves of environmental controversy and influenced the structure and aims of metropolitan government.[70] It is tempting, but erroneous, to conclude that since from the second half of the nineteenth century it became increasingly credible to speak in terms of a generalised 'science' of the environment, so by analogy it may be possible to arrive at an unambiguous evaluation of the effectiveness of anti-pollution measures at that time. Central to the argument presented here is that any analysis of embryonic environmental agencies in nineteenth-century London can be neither separated from larger political and ideological contexts, nor abstracted from the value-laden rhetorics which were so eloquently and persuasively developed.[71]

1   *Hansard* **clxxvii** col 1336.
2   *Ibid*, col 1335.
3   The classic account is Gareth Stedman Jones, *Outcast London: A Study in the Relationship between Classes in Victorian Society* (Oxford, 1971) part I.
4   The writings of H J Dyos have done much to elucidate the process of suburbanisation in the capital. See, in particular, *Victorian Suburb: A Study in the Growth of Camberwell* (Leicester, 1961).
5   The demographic parameters have been authoritatively dealt with in E A Wrigley *Past and Present* **37** (1967) 44–70.
6   There is as yet no full account of this topic but for vivid contextual background see Roy Porter, *English Society in the Eighteenth Century* (1982), chapters 2 and 3. See also W D Rubinstein, 'The End of "Old Corruption" in Britain 1780–1860', *Past and Present* **101** (1983) 55–86.
7   The historical balance-sheet in relation to the Board of Works has been described by David Owen in *The Government of Victorian London 1855–1889: The Metropolitan Board of Works, the Vestries and the City Corporation* edited by Roy MacLeod (Cambridge, MA, 1982) and Francis Sheppard (1971), *London 1808–1870; The Infernal Wen* chapter 7.
8   See Chapter 8 below, and particularly the opening section.
9   *Hansard* **cli** col 431. N Kendall.
10  *Ibid*, col 573.
11  J R L Anderson, *The Upper Thames* (1970) 191–2. This account is, however, over-generous to the City.
12  William A Robson, *The Government and Misgovernment of London* (1939) 131–2.
13  *Hansard* **cxlvi** col 871.
14  *Ibid* **cxlvi** col 876.
15  *Ibid* **cxlvi** cols 867–70. Sir James Graham.
16  *Ibid* **cli** cols 385–6.
17  *Ibid* **cxlvi** col 1130.
18  William A Robson, as note 12 above 131–2; and Fred S Thacker, *The Thames Highway* Vol I: General History (new edition, introduction by Charles Hadfield, 1968) 234 and 239–40.
19  *MBW Report* (1889) 25.
20  *SC Thames Navigation Bill* (1866) Q 3798.
21  *Ibid*, Q 3834.
22  *Ibid*, Q 3848.
23  *Select Committee Thames Conservancy* PP 1863:XII:iv.
24  *SC Thames Navigation Bill* (1866) Q 4512–3.
25  'Memorial from the Vicar and Inhabitants of Barking on the present condition of the River Thames, in consequence of discharge of sewage' PP 1868–9:L:475 *passim*.

26    *MBW Report* PP 1871:LVII:15–16.
27    *Ibid*, 16–17.
28    *Royal Commission on Metropolitan Sewage Discharge* PP 1880:XL:Appendix O.
29    *MBW Report* PP 1881:LXXIX:16–17.
30    *MBW Report* PP 1882·LIX:17–18.
31    *MBW Report* PP 1883:LIX:8–9.
32    *MBW Report* PP 1884:LXVII:7–10.
33    *MBW Report* PP 1887:LXXI:6–8.
34    *Hansard*, fourth series, **xv** col 432.
35    *Ibid*, col 453.
36    *Ibid*, col 440.
37    Thames Conservancy, *Minutes: Parliamentary Committee*, 20 February 1894.
38    Thames Conservancy, *Minutes: River Purification Committee*, 13 April 1896 and 29 November 1897.
39    *Ibid*, 1 November 1897.
40    *Ibid*, 12 December 1897.
41    William A Robson, as note 12 above, 138.
42    *SC Thames Conservancy* (1863) Qs 1324–9. Evidence of B J Sullivan.
43    *Ibid*, Qs 7053–4. Evidence of J Thorp.
44    *Ibid*, Qs 1908–22. Evidence of B J Sullivan.
45    *Ibid*, Q 1465. Evidence of B J Sullivan.
46    *Hansard* **clxix** col 494.
47    G E Walker, *The Thames Conservancy 1857–1957* (1957) 6.
48    *Hansard* **clxxxiv** col 764. Lord Eustace Cecil.
49    *Ibid*, col 763. T Milner Gibson.
50    *SC Thames Navigation* (1866) Q 2405.
51    See Chapter 1 and the works cited in note 39 there.
52    *Hansard* **clxxxiv** col 767.
53    *Rivers Pollution Commission: First Report: River Thames* PP 1866:XXXIII: 32.
54    *Royal Commission on Metropolitan Sewage Discharge* PP 1884–5:XXXI: Appendix DQ. Evidence of Sir Thomas Nelson. See also the letter to the Conservancy Board from its legal advisers: Thames Conservancy, *Minutes*, 28 September 1876.
55    See, for example, Letheby's comments in Thames Conservancy, *Minutes*, 21 November 1870.
56    *General Report of the Conservators of the Thames* PP 1871:LVI:802.
57    *Conservators' Report* PP 1880:LVI:700–1.
58    *Hansard* **cclx** col 134. Baron Henry de Worms, member for Greenwich.
59    *Ibid*, 102.
60    *Conservators' Report* PP 1881:LXXV:896; and G E Walker, note 47 above, 12.
61    *Conservators' Report* PP 1886:LII:766; and *idem* PP 1890:LVIII:834.
62    *Conservators' Report* PP 1889:LX:1012.
63    *Conservators' Report* PP 1859:LXXX:637.
64    Thames Conservancy, *Minutes*, 11 November 1893.

65    Thames Conservancy, *Minutes: River Purification Committee*, 29 October 1894.
66    *Conservators' Report* PP 1896:LXVIII:547.
67    *Conservators' Report* PP 1899:LXXVIII:533.
68    Thames Conservancy, *Minutes: River Purification Committee*, 12 February to 19 November 1894.
69    Thames Conservancy, *Minutes: River Purification Committee*, 3 March 1899.
70    See here for theoretical framework and contemporary parallels Mary Douglas and Aaron Wildavsky, *Risk and Culture: An Essay on the Selection of Technological and Environmental Dangers* (Berkeley, 1982) and John Passmore, *Man's Responsibility for Nature* (1974) chapter 3.
71    This historiographical dimension has been enriched by the work of Quentin Skinner. See, in particular, his 'Meaning and Understanding in the History of Ideas' *History and Theory* **8** (1969) 3–53. Also relevant here is the approach of Gareth Stedman Jones, 'Rethinking Chartism' in *Languages of Class: Studies in English Working Class History 1832–1982* (Cambridge, 1983).

# 8 The Failure of National Legislation

This chapter is concerned with the failure of national legislation to combat the pollution of rivers during the second half of the nineteenth century. This is a topic which requires examination within the context of common law traditions which both predated and outlived the first consolidating statute in this field—the Rivers Pollution Prevention Act of 1876. More specifically, it demands consideration against the back-cloth of Chancery actions and their deployment by and impact upon landowners, manufacturers, boards of health and corporations. It was, in fact, the unsatisfactory nature of Chancery—that recourse to it was available only to the wealthy and that it moved with such ponderous slowness—which convinced reformers that environmental conflicts would be more effectively resolved by means of specific legislation. The national and parliamentary campaign between the mid-1860s and the mid-1870s was characterised by confrontation between industrial and agricultural interests; centralisers and champions of local self-government; and supporters and opponents of the scientific 'standards' laid down by the Rivers Pollution Commissioners.[1] The decisive battle was fought between 1872 and 1875 and was contested by those who were convinced that scientific criteria could be objectively applied to evaluate the quality of manufacturing and sewage waste flowing into rivers, and manufacturing groups representing tanning, paper-making, dyeing and alkali enterprises in Yorkshire and Lancashire, who insisted that stringent implementation was impracticable and a threat to the 'continuation of trade'.

Greater emphasis is given here to the ways in which Disraeli's ministry was first influenced and then effectively outmanoeuvred by industrial interests than to relations between central government and local boards responsible for the disposal of sewage. Once the ideal of scientifically-backed inspection had been discarded and the Government had been persuaded that the Local Government Board must act as a 'buffer' between manufacturers and economically harmful prosecution, the cutting edge of the proposed legislation was blunted both in relation to businessmen and local boards. The opposition of manufacturing groups, in other words, involved so extreme a dilution of the principle of intervention that sanitary authorities, which had originally been conceived as implementers of the act, were now themselves able to demand that they be taken under the protective wing of the Local Government Board.

But we should not expect that environmental legislation should or could have been effective during this period—that would be to underestimate the power of *laissez-faire* and of a dominant ideology which bound reformers as much as supporters of the status quo to a belief in the inherently non-antagonistic 'collaboration' between capital and labour, and the no more than intermittent importance of the public domain.

The most stringent court order against river pollution was an injunction in Chancery, backed up by threat of sequestration. But this was often the culmination of a long period of wrangling, of accusation and counter accusation, with one party failing to adopt satisfactory corrective measures and the other pressing for ever more punitive vetoes. Disputes might drag on for years and, because the law was so notoriously costly, it was only the wealthy who could afford to confront a manufacturer, local board of health, or corporation in Chancery. Cases involving collective bodies could lead to unforeseen complications, with an injunction granted to a landowner which forced a board to change its method of sewage disposal, clearly having an effect on the health and welfare of very large numbers of people; in towns which were temporarily compelled to revert to cesspools and middens, the comfort of those living in the countryside might well be protected but only at the expense of urban populations. In terms of industrial effluent, Chancery decisions tended to favour customary and prescriptive rights. If, in other words, a nuisance had existed for a long time it was likely to be deemed permissible solely because it had existed for a long time. And those who sought to show that an industrialist had forfeited his right to customary immunity usually found the burden of proof exceptionally onerous.

The complexities of Chancery proceedings are well illustrated by the protracted case of *Spokes v Banbury Local Board of Health* heard before Vice-Chancellor Wood in 1865. The Spokes family had lived for several generations at a mill on the river Cherwell just outside Banbury, a town which had taken up the Public Health Act in 1852, and which had constructed an evidently unsatisfactory sewage works by 1858. By this date, the plaintiff said, the Cherwell had become 'an open sewer . . . quite black, abominable in stench, filled with lumps of floating filth, and covered with bubbles of putrefactive gas'. It was also claimed that there had been injury to fish and cattle, to haymakers and to the health of the plaintiff's wife. The board's reply was that it had sought the best means of deodorisation and treatment and that it would try to do better. But the judge was unsympathetic and emphasised that 'the Act of Parliament under which the defendants derived their powers expressly stated nothing therein contained should legalise the committal of any nuisance'.[2]

An injunction was granted and the town was given three months in which to find a way to cleanse its sewage more effectively. When the case next came to court the board explained that it had sought help from the inspectorate at the Local Government Act Office and that a deputation had been sent to Stroud to examine a new ammonia technique which might be suited to Banbury.[3] But there had been no change in the state of the Cherwell. 'The water', it was now claimed, 'had become like ink, evolving a poisonous effluvium to such an extent as to kill the fish, tarnish the silver spoons in the mill, and the watches and the money in the workers' pockets, and produce nausea, sickness, loss of appetite and depression of the nervous and vascular system in the inhabitants on the banks of the river.'[4]

The owner of the mill and his wife had been forced to leave, and a mill worker had apparently taken to the bottle. Finally, between 25 September and 16 October, all work had stopped because of the stench. The judge accordingly sequestrated the town's stone yard, its cart shed, the land on which were located the offending sewage tanks and a municipal refuse heap. It was no excuse, he said, that the board had consulted the Local Government Act Office. 'Take the case', he went on, 'of a man who for his own convenience threws his sewage into his neighbour's yard. When he is ordered by the court not to do so to his neighbour's annoyance and injury, is he to be allowed to come here with a story that he has consulted the most eminent chymical authorities, that no means can be devised for deodorising the filth, and that he cannot help throwing it upon his neighbour's property?'[5]

Throughout the 1860s and 1870s local boards, some energetically, others lethargically, were tackling river pollution under the shadow of injunctions similar to that granted by Vice-Chancellor Wood. When the Sanitary Commission of 1869–71 was taking its evidence Bradford was under an order to cleanse its sewage by 1872;[6] Coventry reported that it was being regularly threatened with an injunction by a riverside proprietor;[7] Norwich was forbidden absolutely to discharge anything at all into its river; and Northampton was involved in injunctions and litigation despite the fact that it was employing precipitation and deodorisation before discharging its sewage.[8] No doubt many towns and, especially, smaller rural sanitary boards were dilatory and may have been spurred on by aggressive legal action. But, as Royston Lambert has pointed out, others were more adventurous and anxious to apply new ideas than investigatory bodies like the Sanitary Commission may have allowed. For a progressive board the militancy of riparian owners was a hindrance rather than an encouragement, especially when, as at Birmingham and elsewhere, more efficient pollution control depended upon the purchase of land from the very landowners who were seeking what might in any case be little more than short-term satisfaction at Chancery.[9]

The plans, for example, of the Nantwich Board, which was placed under an injunction in 1870 not to pollute the River Weaver on pain of a fine of £10 000, were delayed by the unwillingness of several riparian owners to sell. A condition of the order was that no further sewers should be connected to the offending outfall until a new sewage farm had been completed—and there would, of course, be no farm without the land on which it could be constructed. This constraint has been held to have contributed to the epidemics of smallpox which struck the town in 1871 and 1873.[10] Whatever the validity of that contention, agriculturalists' fears were rapidly shown to be unfounded: once sufficient land had been bought, the farm built, and the legal veto lifted, observers noted improvements in the fertility of the fields bordering on the new sewage area. Conflict, both here and elsewhere, between landowners and corporations was rooted in ingrained political, social and cultural animosity: mutually satisfactory solutions were technically feasible but hostility between town and country only belatedly discarded.

The position could be equally tense when two boards became locked in legal battle, since it was always possible that one of the parties to such a conflict might actually distintegrate in the face of an injunction. Even more bewilderingly complex was the fact that there might be an offence but no existing agency to sue for it. In 1866 the Tottenham Board started proceedings against Hornsey because the latter was claimed to be fouling a stretch of water called the Moselle stream. Although few people doubted Hornsey to be in the wrong the action was dropped because there was 'no public body in Hornsey to be made amenable'.[11] Depending, then, on the relative strength or weakness of a local sanitary authority, an injunction in common law might be too lax, too overbearing or blatantly counterproductive. 'Sometimes' Lord Robert Montagu reflected, 'the peccant town has continued to perform its pollutions, disregarding the injunctions of the court, the threats of penalties, and the law of the land. Sometimes . . . the local board, instead of being forced by the injunction to carry out the law, has itself been crushed by the injunction, and no local authority has consented to succeed it.'[12] The overbearing riparian owner or self-important sanitary board might therefore be more seriously harmed than its polluting, but administratively feeble, neighbour; a weak board downstream was preferable to no board at all.

Although, as we have seen, the balance in common law seems to have been tilted in favour of agriculturalists against sanitary authorities, manufacturers frequently gained from the bias towards prescriptive right. In a revealing appeal in Chancery in 1867 an estate owner at Rickmansworth claimed that a papermaker no longer had a right to empty polluting water into the river Chess on the grounds that it was a more noxious fluid than that which had been used by the previous owner. The new material was the notoriously dirty esparto grass but the conclusion of Lord Justice Cairns was that 'it is not sufficient for the plaintiff to show that the defendant uses . . . a raw material different from that formerly employed: he must show, further, a greater amount of pollution and injury arising from the use of this new material: and the onus, of course, of showing this lies on the plaintiff'.[13] The task of 'proving' pollution was considerable, especially when there was a conspiracy between businessmen and a local board, or a tacit agreement among tradespeople to share the costs of litigation and thus hold off the small man who might be contemplating action. John Craven was a mill owner in Keighley, Yorkshire. His pool had been obstructed by industrialists further upstream who had thrown coals into the water-course. He would, he said, find litigation far too expensive because the 'foundry people up the river have entered into a sort of bond that each will bear a proportion of the expense of any litigation that may arise upon the matter, so that whoever has to fight them, fights at a great disadvantage'.[14] S H Gael, a barrister who prepared the first Nuisances Removal Bill and the Metropolis Sewer Consolidation Bill of 1848, confirmed Craven's depiction of Keighley. The water level had risen by ten to fourteen feet, yet the local board had refused to take any action to prevent manufacturers and iron foundry owners from throwing ashes into the river. So intimate was the relationship between the board of health and the industrial interest in this particular locality that any legal intervention would have been tantamount to self-prosecution. 'An authority like a local board of health' Gael warned, 'is one of the most difficult to call to account for not doing its duty that can be conceived.'[15]

The small businessman, then, was exceptionally vulnerable. A supplier of water at Knaresborough, whose family had been in the same trade since the late eighteenth century, told the Rivers Commissioners in 1867 that he had been threatened with legal action because his own locally drawn supplies were being contaminated by sewage flowing all the way down from Harrogate. Should he take action against Harrogate? Or against the local board for not taking action against Harrogate? How could he best defend himself in common law? 'I might perhaps have to go to Chancery and I am too old to go to Chancery' was his resigned reply.[16] Resort to the ancient legal institution was unsatisfactory for all but the wealthy and powerful. An injunction might ensure a temporary restraint but it did little to change the environmental context in which disputes developed or to encourage wrongdoers to take steps to prevent a recurrence. It was a static solution based on concepts of prescriptive property rights in a society in which collective boards were increasingly entrusted by statute to care for the health of great populations. Chancery was intensely conservative at a time when the notion of 'nuisance' was undergoing objective transformation: the simple annoyance had become, for the town-dweller, a grave threat to health. In 1873 the Lord Chancellor, Lord Selborne, held that 'it would not do to stop manufactures and works off hand, especially in areas where legal rights to discharge refuse matters into streams existed, so far as private persons were concerned, and where, according to the existing law, there was no public nuisance.' But later in the same speech he noted that 'it might be that in obeying it [an injunction] as best they could the authorities would do something, which would create another nuisance quite as great as that for the putting down of which the aid of the Court of Chancery had been invoked, and then another suit by other parties became necessary. In all cases which arose in connection with the pollution of rivers it had to be considered whether any means could be found for abating the mischief.'[17] This was the context within which reformers and parliamentarians began to press for new laws against the pollution of rivers.

There had been growing pressure since the mid-1850s for national legislation. An attempt had been made to fine those who fouled waterways with manufacturing effluent under the Nuisance Removal Act of 1855.[18] Then, in 1861, the issue was taken up by the Fisheries Preservation Society, which drafted the Salmon Fishery Act of that year, but the anti-pollution clauses proved largely inoperative.[19] Three years later a Parliamentary report was still insisting on the 'important object of completely freeing the entire basins from pollution . . . by general legislation'.[20] And in 1864, also, a joint letter to Palmerston from the Sanitary Associations and the Fisheries Protection Association led to the appointment of the Rivers Pollution Commission.[21] By now the problem was claiming the attention of scientists and social reformers, parliamentarians and pressure groups.

When he introduced the River Water Protection Bill of 1865 Lord Robert Montagu made use of arguments which had been current since the 1850s. Pollution, he claimed, had come within a hair's-breadth of destroying the freshwater fishing industry; it was a threat to livestock in nearly every county; and it had reduced the profits of manufacturing activities which depended on regular supplies of clean water.[22] But there were deeper and more insidious forces at work. The connections between 'bad water, bad health and drunkenness' increased the costs of hospital treatment, eroded social discipline, and led to the 'unproductive maintenance of widows and orphans'.[23] Equally insupportable was the vast waste of sewage: this must be returned to the land to nurture higher domestic yields and reduce reliance on imports of foreign grain. 'A measure of human manure', Montagu asserted, 'made a measure of corn.'[24] Those who opposed the bill disagreed more with the proposed remedies than the diagnosis. The measure, it was said, 'gave very large powers to an inspectorate who would, also, need to be superhuman to carry out their duties'.[25] Montagu's boards, another critic complained, would be armed with 'powers infinitely more stringent than those given to railway companies' and they would invade 'the land of every one who had the misfortune to have a stream flowing through it'.[26] The manufacturing interest was equally unsympathetic. John Bright spoke in terms of a 'stoppage' of industry.[27] A Yorkshire member foresaw the closure of 'all the mills of Yorkshire and Lancashire'.[28] W E Forster was willing to admit that 'the evils referred to ought to be removed when they could be prevented by manufacturers at such a reasonable cost as would not stop their trade'. But even though the people in the northern manufacturing towns might be uncomfortable '. . . it should be recollected that were it not for manufactures there would be no people at all'.[29] Warning had been given—industrialists and their supporters would fight tooth-and-nail against any legislation which cut deeply into profits or gave undue power to 'comprehensive' boards to reprimand or fine.

Montagu's bill was voted out and when, in 1872, James Stansfeld at the Local Government Board, attempted to insert clauses against river pollution into his Public Health Act, the opposition was more militant and better coordinated. There were those like Lyon Playfair, who insisted that a degree of control was in the best interests of all manufacturers and of industrial regions as geographical and environmental wholes: it was reprehensible that the less affluent firm could be destroyed by the activities of the really large concern. Those higher up the river, Playfair pointed out, 'profess to believe that the complaints of those lower than themselves in the stream are utterly unreasonable. They are the wolves who foul and muddy the stream, and who growl at the innocent lambs below, if they raise their meek eyes in complaint of their misdoings.'[30] Others, like Sir Charles Adderley, argued that 'the more Parliament forced them [the manufacturers] to try all means, the more men had found out that there was a use for almost all supposed refuse, and that in moving an injury to their neighbours, they were actually profiting themselves'.[31] And Lord Robert Montagu continued his plea for strict inspection and control in every watershed in the land.[32] But members from Lancashire and Yorkshire were now opposed not only to the anti-pollution clauses but also to the chemical standards proposed by the Rivers Commissioners for evaluating the insalubrity of a given sample of water. Joshua Fielden, representing the West Riding, launched a scathing attack on these scientific clauses, arguing that they 'would have the effect of shutting up and stopping all the manufacturies' and place too much power in the hands of 'busy-body' local authorities. Far better, he said, to adopt a more flexible scheme which would determine the needs of individual regions and rivers and empower the Local Government Board to protect individual businessmen against frivolous prosecution. 'The evils of pollution' must 'be weighed against the value of our manufactories.'[33] It would later emerge that Stansfeld himself had had no great commitment to or understanding of the chemical standards.[34] There had, moreover, been a tacit understanding with the extremist *laissez-faire* Alkali Association that the anti-pollution clauses constituted little more than a framework for discussion. If adopted, they would be stringently applied wherever river water was needed for public supplies but only loosely interpreted if there was any interference with a productive process which might be 'ruinous to large interests'. It was also said that Stansfeld had readily agreed to the manufacturers' insistence that the Local Government Board should 'protect' industrialists against precipitate prosecution by prejudiced local boards.[35] It is hardly surprising, in the light of all this, that the clauses were omitted from the final act or that manufacturers were now convinced that it would be impossible for any government to pass effective anti-pollution legislation.

However credible they might initially seem in relation to sewage pollution, the next three attempts at legislation—Shaftesbury's bill of 1873, Salisbury's government measure of 1875, and the Rivers Pollution Prevention Act of 1876—were in fact fatally undermined by the growing confidence and influence of manufacturing pressure groups. In 1873 the Fisheries Preservation Association drafted a bill based on the clauses which had been omitted from the Public Health Act of the previous year. It was introduced into the Lords by Shaftesbury and then referred to a select committee which, despite being assailed by a dispro-portionate number of witnesses opposed both to the standards and to virtually every kind of rigorous control, reported in its favour. The fact that the bill was later withdrawn in the Commons[36] was less significant than what it revealed both of the strength of the industrial lobby and its growing ability to influence and actually shape the legislative process. Lyon Playfair and Edward Frankland favoured the enforcement of lenient criteria for the objective measurement of trade effluent but they were outmanoeuvred by trade representatives and their advisers. Playfair pointed out that 'under the present state of the pollution law, no manufacturer is ever sure that, when he has expended a good deal of money in making a manufacture, somebody will not come above him and pollute the water, and destroy the capital which he himself has invested'.[37] In terms of the operation of any new act, he advocated an arrangement which would have enabled a central board either to put pressure on or actually replace a local authority which failed to take action against a persistent polluter—and here he drew an analogy with the powers vested in government to set up school boards in areas in which the voluntary societies failed to make adequate provision for elementary education.[38] Frankland also supported the creation of a central body empowered to act if either a local board or an individual were intolerably dilatory. He was ready to admit that manufacturers might be confronted by technical problems which might make it difficult to deal rapidly with noxious effluents but he could not accept this as an adequate reason to abandon the application of objective and universal chemical criteria. He pointed out that the standards were in fact relatively undemanding and had been devised precisely in order to persuade businessmen that they could be easily and cheaply complied with—there was certainly no intention that water which had been taken from a river, and used in a productive process, must be purified up to a point at which it would be fit for human consumption. He acknowledged that many manufacturers were convinced that an insistence on the treatment of waste water would place them under the control of patent-holders who were able to charge exorbitant fees for the use of new devices. But this, he argued, was to underestimate the speed of cost-cutting innovation, with large numbers of inventors vying with one another in an expanding market. Frankland was sympathetic to the complaint that crowded inner-city conditions, shortage of space, and high land values made it exceptionally difficult to install the necessary plant and suggested that in circumstances such as these pollution control should be seen as a communal responsibility, with businessmen merely paying 'a little more to the rates if necessary'. Like Playfair, Frankland was convinced that 'clean production' would prove to be more profitable production, and he also lent his support to the idea that manufacturers who observed the standards should be granted exemption from common law actions.[39]

William Crookes, for the manufacturers, insisted that universal chemical standards were both unfair and unworkable. The emphasis, he contended, should be on the individual river and its precise locality, rather than the quality of effluent flowing into it. 'It would only be fair', he said, 'that a manufacturer, drawing pure water near the course of a river, and then polluting that river throughout its whole course, should be judged by a severer standard than if he were to do exactly the same thing 50 miles below, where the river water, as he received it, was already polluted.'[40] In administrative terms, Crookes went on, 'there would be such great difficulties in carrying out a hard and fast test like these all over England, with rivers of varying amounts of original purity, and with so many manufactures, that I really do not see how the rivers can be practically purified by such an Act of Parliament at all'.[41] What was required was the type of consultative relationship that had grown up since 1863 between government and the alkali industry, as well as a strong central authority which would protect industry against unnecessary prosecution.[42] Spokesmen for the trades which would have been affected by the application of the standards testified aggressively to the severity of the difficulties with which they would be confronted. The Alkali Association objected to 'being compelled by law to adopt any process which is the subject of a patent. You are in the hands of the patentee, who might refuse licences, or impose any terms he chose': it was even conceivable that the measure might 'extinguish the manufacture'.[43] A witness for the tanners of Leeds claimed that the legislation could bring the industry to a halt in that town. 'I am of the opinion', he went on, 'that any act of legislation ought to define the manner of purification, and specify processes, and ought not to make penal a result of commercial industry, the prevention of which it cannot demonstrate to be practicable at a reasonable and fair expense. . . . The evil now existing is much less than the serious evil which would arise from the destruction of the trade in the district.'[44] The dyeing industry protested that lack of space in densely populated inner-city areas made effective water treatment impossible.[45] And a spokesman for the colliers of Northumberland and Durham pointed to the 'enormity of the interests' that depended upon them. 'We should not', he told the committee, 'like to be in the hands of the magistrates day after day, with fines imposed upon us. We should not like the payment of the fines: and I think we should still less like to be in antagonism with the law.'[46]

Powerful members of the Lords, and particularly Salisbury, were clearly impressed by the employers' case. Salisbury believed that the proposed system of progressive fines would be unfair both to agricultural and manufacturing interests; that the scientific principles underlying the measure must be tempered with 'practicality'; and that many rivers, as William Crookes had pointed out, were more polluted than the liquids which would be forbidden to flow into them. As the bill began to lose impetus, Salisbury pressed for precisely that 'flexibility' and 'relativity'—with the individual river rather than the potentially polluting effluent determining the shape of any legislative panacea—which those opposed to the measure, industrialists and chemists alike, had so uncompromisingly articulated.[47] This interpretation is supported by the fact that the government bill, introduced by Salisbury in 1875, omitted all mention of the Rivers Commission's standards and increased the power of the Local Government Board to veto economically damaging prosecutions. George Sclater-Booth, who was President of the Local Government Board, and whose department had drawn up the bill, professed himself committed to a firm measure to inhibit the 'polluting interest'. What emerged, however, was a basically moderate bill, which had been examined, confidentially, by Frankland, Robert Rawlinson and Angus Smith, chief inspector of the Alkali Administration. 'Mr Rawlinson', Sclater-Booth assured the Cabinet, 'has probably more complete knowledge on the subject than any other living man and his views as to what is possible to "pass" have been always on the side of extreme moderation.'[48] But Salisbury was unconvinced. He admitted that, in theory, any effluent could be purified. But many manufacturers lacked adequate and appropriate technology and the expense involved might be 'equivalent to ruin'. The bill demanded that a manufacturer must use the 'best practicable and available means' but, Salisbury argued, a county court judge who would be required to make a decision on this issue, might claim that 'as long as a man has any money left, a course is not impracticable on account of its expense. It may be harsh, unreasonable, ruinous; but it is always practicable to make a man pay up to the last farthing that he has got.'[49] It might also be necessary, Salisbury believed, for a judge who wished to evaluate what was 'reasonable' to inquire not only into the cost of new works but also into the 'private fortune' of the businessman who would be required to construct them. He was now convinced that the power that was to be given to the county courts must be greatly reduced. 'When one County Court Judge', he told the Cabinet, 'has the power, without straining the law, of ruining a large part of Leeds, and another may spare Manchester altogether, it is obvious that the temper of the County Court Judge will become a more serious element in the prosperity of a town than the proximity of a railway or a port.'[50] But what were the alternatives? Salisbury cited the example of the Railway Commission—a body which consisted of 'more than one man', commanded public confidence, and had jurisdiction over a 'considerable area'. He had conferred, he said, 'with a great number of persons interested, from all parts of the country, and I am convinced that, without the creation of a strong tribunal to exercise powers which are necessarily arbitrary and ill-defined, the Bill is not defensible in argument'.[51]

The bill, which had been weakened in the Lords, failed to reach the committee stage in the Commons.[52] Sclater-Booth had been outflanked and Salisbury now moved his growing authority behind an even more diluted Rivers Pollution Prevention Bill in 1876. The shape of this measure had now become almost wholly predetermined. There were strong indications, it was suggested, that the Government had simply drawn up a framework and then invited the business interest to fill in the details.[53] During the committee stage in the Commons, Playfair argued that the bill had the support of every manufacturing pressure group because it was in fact 'in the interest of both the traders and the manufacturers'.[54] Sclater-Booth made little effort to deny these accusations and stressed that 'the enormous capital at stake was not to be lightly dealt with'.[55] Playfair sustained his lonely opposition. 'The Bill is so altered', he protested, 'that public interests seemed to have vanished altogether in the background, and the interests of the manufacturers are pushed into prominence in every clause. . . . Take, for example, the eighth clause, which forbids even the Local Government Board to give consent to proceedings under the Act, if it will inflict injury on the interests of any manufacturing industry. In the first edition that prohibition was not there at all, then in the second edition the Bill threw the mantle of protection over the textile industries, and now by the third edition the Local Government Board is to protect all industries whatever against the public rights to have rivers free from pollution. No wonder that manufacturers highly approve this Bill.'[56] Deputations called on Sclater-Booth to the very last: but, given the persistent pressure which had been so professionally applied during the last five years, a body like the Sanitary Institute could hardly be expected to carry as much weight as the Papermakers' Association.[57]

This first national legislative attempt to reduce river pollution was divided into four parts covering manufacturing solids, solid and liquid sewage, manufacturing liquid, and mining solids. Provided that he availed himself of the 'best practicable and available means' no manufacturer could be prosecuted under the Act. But even this nebulous phrase was weakened by a clause which allowed pollution to continue in a channel leading to a river if it 'had been constructed or was in process of construction' at the date of the passing of the Act. As a critic pointed out some years later, this allowed 'a dozen manufacturers who would each be liable to the prohibitions contained in the act if they construct new outlets for their refuse . . . to pollute the river without fear of any real interference if they made common use of a drain that happened to be in use by one of them in 1876'.[58] Complaints of pollution against manufacturers and mine owners could only be followed up by a vested sanitary authority, and only then if the authority had been given explicit permission to prosecute by the Local Government Board. Manufacturers were further cushioned by a provision which gave them a right to make their own case to the sanitary authority before any legal proceedings were allowed to begin. Complaints by individuals, or groups of individuals, were expressly discouraged and this ensured that many cases continued to find their way to Chancery. Manufacturers were granted two further screens against what was considered to be unreasonable legal harrassment. First, it was possible to apply to an inspector for a 'good conduct' certificate, of two years' duration, confirming that a business was in fact making use of the 'best practicable means'. Secondly, the state of employment in an area was always to be considered before the Local Government Board would allow an action to be taken by a sanitary authority. 'In giving or witholding their consent', explained John Lambert, Secretary to the Local Government Board, 'the Board are to have regard to the industrial interests involved in the case, and to the circumstances and requirements of the locality; and they are prohibited from giving their consent to proceedings by the Sanitary Authority of any district which is the seat of any manufacturing industry, unless they are satisfied . . . that means for rendering harmless the liquid refuses from the processes of such manufactures are reasonably practicable under all the circumstances of the case, and no material injury will be inflicted by such proceedings on the interests of such industry.'[59] As for the less controversial clauses against sewage pollution, these ensured that sanitary authorities, like manufacturers, were to be carefully shielded by the Board and were also eligible for 'good conduct' certificates.

In a rare lapse of judgment Royston Lambert described the legislation of 1876 as the 'first effective and consolidating Rivers Pollution measure'[60], while George Kitson Clark accounted for its shortcomings in terms of the unavailability of the 'right scientific knowledge'.[61] Other historians have been less generous[62]; and many late-nineteenth-century observers viewed the measure as one which had been shaped by and in the interests of the manufacturing class. As for the control of sewage pollution in the immediate aftermath of the Act, as many as 19 of the 53 proceedings sanctioned by the Local Government Board between 1880 and 1885 were taken out against 'local Boards, Corporations, or Sanitary Authorities—the very bodies to whom the exclusive carrying out of the Act is entrusted'.[63] In 1883 Angus Smith, then inspector of rivers as well as of alkali works, attempted to defend the Act but admitted that too little had been achieved and that the measure lacked public credibility.[64] And in 1886 Sclater-Booth (by then Lord Basing) acknowledged the grave weaknesses of the Act and advocated a strengthening of its provisions 'so as to secure greater activity and vigilance'.[65] 'The local sanitary authority', another critic noted ironically, 'who were generally the chief offenders, were entrusted with the power of enforcing the act.'[66]

Until the end of the century and beyond, any suggestions that the law against river pollution should be either revised or wholly redrawn were countered by an insistence that such legislation would cause irreparable damage to the economy. In a remarkably frank interview in 1887, Lord Stanley, the President of the Board of Trade, had apparently admitted that 'in the present depressed state of trade it was no use recommending to the Government that any measures should be proposed which would in any way restrict the present licence, if he might use the term, which was accorded to mar..ufacturers'.[67] Only rarely is the unanimity of state and industry under capitalism so transparently revealed. By the early 1890s, though, a number of factors such as alarmingly high levels of river pollution, continuing dissatisfaction with the legislation of 1876, and administrative changes connected with the Local Government Act in 1888, had stimulated renewed parliamentary activity. In 1892 the county councils of Lancashire and Cheshire set up a joint committee of eight county boroughs in an attempt to halt the dangerous and escalating deterioration of the Mersey and Irwell. By means of a local act which, according to C T Ritchie, President of the Local Government Board, swept away 'the whole protection given to manufacturers and others', the committee became the major prosecuting body in its region. In 1893 a short but important bill made it compulsory for every local authority in the country to apply the 'best practicable and available means' of treatment to sewage before allowing it to enter a river.[68] The same year witnessed the creation of the West Riding Rivers Board, a body empowered to prohibit sewage pollution and, with the agreement of the Local Government Board, to take action against manufacturing effluent. During the next 10 years urban districts in the region without sewage works declined from 81 to 39 and the corresponding reduction for non-county boroughs was from six to one. Of 28 major trade effluents entering a minor river in the area in 1896 only 10 were being treated: by 1902 the figure was 24 out of 30.[69] In 1898 English boroughs, like county councils, were empowered to set up joint anti-pollution committees.[70]

But, more than 30 years later, an official inquiry revealed that both the law and the administration of the law relating to river pollution were in a state of dire confusion. The linchpin was still the Rivers Pollution Prevention Act of 1876 but action could also be taken under statutes ranging from the Waterworks Clauses Act of 1847 to the Diseases of Animals Act of 1894. In theory, enforcing bodies included every town, urban, county and rural district council; joint committees with special responsibilities for the Mersey and Irwell, the upper waters of the Tame and every river in the West Riding of Yorkshire; and designated fishery boards. 'Unity of control', the reforming slogan of the 1860s and 1870s, had not been achieved.[71] It is scarcely surprising, therefore, that there was continuing and widespread recourse to common law expedients. 'Nearly all legal proceedings', an experienced observer commented in 1887, 'against parties polluting streams had been taken at common law and not under the statutes.'[72] Some of these, such as the tortuous conflict between the Lea Conservancy Board and the Tottenham Board, which began in 1869 and ended in 1886, were lengthy, impassioned, and only fully intelligible to those well versed in the law.[73] Others were identical to those which have been described in the first part of this chapter. 'Some riparian owner', it was reported in 1893, 'finding the stench unendurable, or some water company or manufacturer drawing a tainted supply from the polluted river brings an action. After some eighteen months he obtains an injunction, the effect of which is suspended again and again in order to give the offending Board of Health plenty of time to examine the competing chemical methods, to buy a small piece of land, to get the sanction of the Local Government Boad to borrow money to put up works.'[74] Despite Edward Frankland's recommendation to the contrary in 1873, a businessman who had fulfilled the requirements of the legislation of 1876 and convinced the Local Government Board that he was making use of the best available means to purify an effluent could still be prosecuted at common law for creating a nuisance.[75] But we should probably feel less concern for the manufacturers of Edwardian Britain than for the great majority of citizens who were still subjected to appallingly high levels of environmental pollution and who were quite incapable of finding the money—anything up to £8000—that a successful suit might demand.[76]

Recent research has pointed to discrepancies and lags between bodies of law at a given historical moment and the dominant class and economic interests which, in the final analysis, the law reflects and reinforces.[77] And it can clearly be seen that exceptionally subtle arguments would need to be deployed to demonstrate that measures framed, for example, to protect potential Irish emigrants from being herded into ships which habitually sank half-way across the Atlantic, or legislation controlling abuses in private mad-houses, constituted *direct* instruments of class domination.[78] What is emphatically not asserted here is that such laws were neutral or motivated wholly out of humanitarianism[79]: the warning, rather, is against modes of historical analysis which *reduce* legal systems unproblematically to systems of class control.

But the evidence presented in this chapter surely indicates that historians have little need to tread lightly when attempting to account for the failure of national legislation against the pollution of rivers during the second half of the nineteenth century. Effectively devised and enforced, such measures would have trespassed in an unthinkable manner on the rights of property and the rights of capital. All this was repeated over and over again, with little subtlety or apology, by those, mainly from the manufacturing districts of Yorkshire and Lancashire, who derided scientific standards and, following Salisbury's 'conversion', successfully dictated terms to Disraeli's government. The threat of non-compliance was always present and there can be little doubt that it was taken seriously. But industrial pressure groups did not always seek either to ignore pollution or to deal with it without reference either to the law or to legal precedent. In this respect Chancery might, on occasion, be well enough adapted to conflicts between wealthy individuals, and between individuals and boards. Predominantly precapitalist in origin and charisma, costly, and slow-moving, the institution so frighteningly depicted by Dickens in *Bleak House* certainly operated with formidable unpredictability. But it only rarely forced a manufacturer to adopt the 'best available' or the 'cleanest' type of productive process. A classically eighteenth century preoccupation with 'nuisance' rather than 'pollution' ensured that explicitly technological aspects of complex environmental confrontations were given only minimal attention. Corporate boards, on the other hand, could be harshly treated, with sewage pollution being conceived as constituting a venerable and unambiguous 'interference' committed by an 'overbearing' corporate entity against the rights of the individual. Private property, according to this view of the world, was sacrosanct, while a board which failed to dispose of its rubbish without offence threatened not only the health of the 'public' but even more fundamentally, the well-being of the individuals who made up that 'public'.

The plight of social reformers and scientists who advocated specific national legislation against river pollution in a society dominated by an ideology as deeply embedded as this was a truly desperate one. It was, of course, possible to champion outright interventionism. 'We learned' the Rivers Commissioners reported in 1874, 'from Mr H E Taylor of Aberystwyth that the mining industry . . . pays no less than £80,000 a year in wages and £20,000 a year in royalties, figures indicating an industry of such importance that it might be dangerous to interfere with its property by penalties. We should, however, rather infer from them that an industry of such magnitude is well able to pay for whatever injury it inflicts.'[80] But there were always disturbing exceptions and contradictions. 'Where the landowner's royalty', the Commissioners stated in the same year, 'and the interest of the capitalist—the interests in short of the whole resident population—are dependent on the maintenance of the river channel in its present character as an apparatus for the recovery of mining produce, it would be the utmost pedantry to insist upon such remedies for river pollution being adopted at the great mines above as would destroy the whole of the vast industry below.'[81] On the face of it, this was reasonable enough; but it was precisely what was most 'reasonable' which could be most easily turned against the cause of reform by those deeply opposed to it. Even when they believed themselves to be acting from a position of genuine independence, those who advocated national legislation against river pollution often found themselves tacitly supporting the presuppositions and legitimations of the dominant ideology.

1    A definitive formulation of the 'standards' may be found in *Rivers Pollution Commission: Sixth Report* (1874) 51. No liquid containing three parts by weight of 'dry mineral' in 100 000 parts of water was to be passed into a watercourse. The comparable scale for dry 'organic' matter was to be one part. There were also restrictions to two parts per 100 000 parts of water for all metals except calcium, magnesium, potassium and sodium and very strict maxima for arsenic, chlorine, sulphuric acid and alkalis. As a more general criterion, any liquid which exhibited a 'distinct colour' when a stratum one inch deep was placed in a white porcelain or earthenware vessel would be forbidden to be passed into a river.

2    *The Times* 2 March 1865.

3    *Ibid*, 27 November 1865.

4    *Ibid*.

5    *Ibid*.

6    *Royal Commission on the Sanitary Laws: Second Report* PP 1874:XXXI:662.

7    *Ibid*, 670.

8    *Ibid*, 690.

9    The extremely lengthy conflict over river pollution and the disposal of sewage in Birmingham may be reconstructed from E P Hennock, *Fit and Proper Persons: Ideal and Reality in Nineteenth Century Urban Government* (1973) 107–11 and W S Childe-Pemberton, *Life of Lord Norton* (1909) 143 and 283.

10    W H Chaloner, *The Social and Economic Development of Crewe* (1950) 126–31.

11    *Rivers Pollution Commission: Second Report* (1867) 14.

12    'Paper by the Rt Hon Lord Robert Montagu MP on Watershed Boards . . .' PP 1876:LX:616

13    *Rivers Pollution Commission: Second Report* (1867) Appendix VIII, 307.

14    *Select Committee on the Sewage of the Metropolis* PP 1864:XIV:Q 4292.

15    *SC Sewage Metropolis* (1864) Q 3281.

16    *Rivers Pollution Commission* (1867) Qs 5730–3.

17    *Hansard* **ccxvi** col 6.

18    C E Saunders, 'Legislation for the Purification of Rivers and its Failures', *Trans. Soc. Med. Off. Health* (1886–7) 76.

19   *Ibid*, 77.
20   *Ibid*.
21   'Letter to Treasury, 4th March, by President of Sanitary Association, and Vice President of Fisheries Protection Association on pollution of streams' PP 1864:L:327.
22   *Hansard* **clxxvii**, col 1312 *passim*.
23   *Ibid*, col 1318.
24   *Ibid*, cols 1322 and 1326.
25   *Ibid*, cols 1331–5. Sir George Grey, Home Secretary.
26   *Ibid*, col 1352. Sir Francis Goldsmid, member for Reading.
27   *Ibid*, col 1337.
28   *Ibid*, col 1344. Lt-Col H Edwards, member for Beverley.
29   *Ibid*, col 1349.
30   *Hansard* **ccxx** col 860.
31   *Ibid*, col 873.
32   *Hansard* **ccxii** cols 1256–7.
33   *Ibid*, cols 852–3.
34   *Hansard* **ccx** cols 883 and 1813. See also *Select Committee (HL) Pollution of Rivers Bill* PP 1873:IX:Q 271. Evidence of John Botterill for the dyeing trade of Leeds.
35   *SC (HL) Pollution of Rivers Bill* (1873) Q 206.
36   C E Saunders, (as note 18 above) 77–8.
37   *SC (HL) Pollution of Rivers Bill* (1873) Q 77.
38   *Ibid*, Q 80.
39   *Ibid*, Qs 122–75 and 462–505.
40   *Ibid*, Q 394.
41   *Ibid*, Q 427.
42   *Ibid*, Q 437. On the alkali inspectorate and its 'consultative' approach see Roy M MacLeod, *Victorian Studies* **9** (1965) 85–112.
43   *SC (HL) Pollution of Rivers Bill* (1873) Qs 191, 197 and 205. Evidence of James Cochran Stevenson.
44   *Ibid*, Qs 250 and 257–8. Evidence of Richard Nickols.
45   *Ibid*, Q 270. Evidence of John Botterill.
46   *Ibid*, Q 303. Evidence of John Burdon Sanderson.
47   *Hansard* **ccxvi** cols 4 and 5; and *idem* **ccxvii** cols 1161–2.
48   PRO 30/6/72 'Cabinet Memoranda Miscellaneous 1874–78', 'Rivers Pollution Bill', 94.
49   *Ibid*, 97.
50   *Ibid*, 98.
51   *Ibid*, 99.
52   *Hansard* **ccxxv** cols 428–32 and 770–80. Speeches by Salisbury.
53   *Ibid*, **ccxxx** col 1676. L L Dillwyn, member for Swansea.
54   *Ibid*, col 1877.
55   *Hansard* **ccxxxi** col 283.
56   *Ibid*, col 282.
57   *The Times* 14 July 1876 and 2 August 1876.
58   *Public Health* **i** (1888–9) 306.
59   *Report of the Local Government Board* PP 1878:XXXVII:175.

60    R Lambert, *Sir John Simon, 1816–1904, and English Social Administration* (1963) 574.
61    George Kitson-Clark, *An Expanding Society: Britain 1830–1900* (Cambridge, 1967) 158.
62    See the comments of Paul Smith, *Disraelian Conservatism and Social Reform* (1967) 258–60; and the ambivalent verdict of Robert Blake, *Disraeli* (1966) 553.
63    C E Saunders (as note 18 above) 81–3.
64    Angus Smith, *Trans. San. Inst. GB* **v** (1883–4) 334.
65    Lord Basing, *Trans. San. Inst. GB* **ix** (1887–8) 71.
66    Lamorock Flower, 'On the Fouling of Streams', *Trans. San. Inst. GB* **ix** (1887–8) 265.
67    Discussion of paper cited in note 66 above, 268.
68    *Hansard* **xi** cols 1283–4. Lord Monkswell. For a discussion of the regional development of administrative intervention at this time see T Richards, 'River Pollution Control in Industrial Lancashire, 1848–1939' *PhD Thesis* (University of Lancaster, 1982).
69    Charles Milnes-Gaskell, 'On the Pollution of Our Rivers', *Nineteenth Century* (July 1903) 93–4.
70    *Hansard* **liv** cols 1149–50. Sir John Dorington, member for Tewkesbury.
71    PRO HLG/50/26 *Joint Advisory Committee on River Pollution: First Report* (Draft) (1928) 2–3.
72    L L Macassey in *Trans. San. Inst. GB* **ix** (1887–8) 272.
73    *Select Committee on Rivers Pollution (Lee)* PP 1886:IX:Qs 117–20.
74    Frank Spence, 'How to Stop River Pollution', *Contemporary Review* (September 1893) 425.
75    A G Leigh, 'Manufacturers and the Rivers Pollution Prevention Acts', *J. San. Inst. GB* **xxiv** (1903) 487.
76    William Ramsay, 'Rivers Pollution', *J. R. San. Inst.* **xxix** (1908) 135.
77    See the masterly analysis in E P Thompson, *Whigs and Hunters* (1975).
78    The examples are taken from Oliver MacDonagh, *A Pattern of Government Growth: The Passenger Acts and their Enforcement 1800–1860* (1961) and William Ll Parry-Jones, *The Trade in Lunacy: A Study of Private Madhouses in England in the Eighteenth and Nineteenth Centuries* (1972).
79    On this point see the incisive interpretation of Michael Ignatieff, *A Just Measure of Pain: The Penitentiary in the Industrial Revolution 1750–1850* (1978).
80    *Rivers Pollution Commission: Sixth Report* (1874) 29.
81    *Ibid*, 69.

# 9  Conclusions

In demographic terms, London was a greatly different city in 1900 from what it had been in 1850. Within the inner region indicated on the map on page 71 aggregate population had risen from just over two and a quarter million in 1850 to four and a half million at the turn of the twentieth century. This represented a very substantial increase but the rate of growth had been declining since the 1880s. During the 1870s the population in this area was rising at just over 1.7 per cent a year; during the 1880s the figure was just over one per cent; and by the 1890s there had been a reduction to around 0.7 per cent. An analysis of demographic change within individual registration districts reveals that, from the 1860s onwards, inner-city areas were experiencing a decline in absolute terms. About 112 000 people had lived in the City of London in 1850; by 1900 this had fallen to just under 30 000. Nearly every other district belonging to the ancient core underwent a similar experience, precipitated by massive slum clearance, the commercialisation of building land, the construction of major railway termini and the burgeoning growth of suburbia. By the first decade of the twentieth century a very large expanse of London, bounded to the south by Lambeth, to the west by North Kensington, to the north by Islington and to the east by Poplar, was losing its residential community at a rate varying from between 0.1 and 2.7 per cent a year, with only Wandsworth and Lewisham continuing to make gains.

The outer suburban ring went through a quite different demographic process, with population rising from about 250 000 in 1850 to just over two million in 1900. The rate of growth held steady at around five per cent a year throughout this period, or just under four times the average recorded for the rest of England and Wales. By the later nineteenth century migrants were consistently more heavily represented in the suburbs than in the inner-city districts: by the early 1880s, about 40 per cent of the inhabitants of 'outer' London had not been born there, whereas the corresponding figure for the inner areas was less than

177

30 per cent. Despite the continuing importance of Irish immigration, which was still running at around 2000 a year in the 1880s, natural increase had long been the major determinant of population growth in London as a whole. The very large as well as socially and politically destabilising in-migrations of the 1830s and 1840s were now a thing of the past and by the later nineteenth century newcomers were probably also less likely to bring serious infections in with them than had been the case two generations earlier. Once settled, they lived in an environment substantially healthier than that endured by their predecessors in the 1840s and after. During the cotton famine of the early 1860s a medical officer reported that an Irishman had made his way from the north west, through the midlands, to the East End only to perish, soon after his arrival, from starvation. No comparable cases were recorded 30 years later.

In some respects, then, those who took an optimistic view of levels of health and environment in late-nineteenth-century London were justified in doing so. Cholera, which decimated the population of Hamburg in 1892, appeared in the English capital for the last time in 1866. Typhoid—year in, year out, a much more demographically significant disease than cholera—was gradually brought under control by means of improvements to the water supply, more effective surveillance of milk and foodstuffs, and increasingly sophisticated epidemiological follow-up of cases attributable to infection via immune carriers. From the later 1880s hospitalisation of patients in institutions administered by the Metropolitan Asylums Board also played a significant role in restricting the scale of epidemics. But typhoid persisted into the twentieth century, claiming victims who either died from it or lived the remainder of their lives debilitated by its after-effects. These sudden eruptions troubled medical officers and others concerned with the prevention of disease in London. Every conceivable mode of transmission was analysed at metropolitan and district level, but often with only minimal success. From what is now known about the aetiology of the disease it is likely that these 'residual' outbreaks were probably attributable to a still less than completely safe water supply and would only be eradicated following the adoption of chlorination under the Metropolitan Water Board in 1915.

If water remained problematic in relation to typhoid during the late nineteenth and early twentieth centuries, its role in the causation and prevention of diarrhoea was altogether more complex. But two factors, the continuing restriction of constant supply in the poorest areas of the city, and intermittent impurities in water consumed and used for domestic purposes, when combined with information on possible changes in the natural history of the disease, are central to an understanding of why it was that diarrhoea continued to kill so many infants until the beginning of the First World War.

In terms both of quantity and quality London's water supply at the very end of the nineteenth century was much less dependable than its more effusive supporters were ready to admit. We still know too little about the internal, day-to-day workings of the companies during the period as a whole but commitment to a capitalistic rather than collectivist interpretation of their responsibility to the capital and its inhabitants was repeatedly asserted. Sceptical towards, and at times dismissive of, scientific and medical knowledge, a majority of the concerns resisted innovation in the sphere of water purification and were still, as late as the 1890s, opposing reformist demands that technical homogeneity was both desirable and feasible. Governmental control over the companies was exceptionally weak and poorly defined, and the right of individuals or groups of individuals to complain about the quality of supplies was so hedged around by bureaucracy as to be little more than notional. In circumstances such as these it is less surprising that the capital was still sporadically afflicted by water-transmitted infection in the later nineteenth century than that such outbreaks were both relatively infrequent and spatially restricted. The demographic and epidemiological record in London at this time lends credence to the view that, in bacteriological terms, very little separates a generally safe from a disastrously dangerous system of public supply.

The absence of cholera, and the arguments of the optimists, ensured that the Thames was not abandoned as a source of supply. London had weathered two 'plague-like' onslaughts of cholera in the mid-nineteenth century, with administrative and engineering remedies being dangerously delayed. But the construction of the main drainage system, together with the much-criticised anti-pollution measures of the Conservancy Board, contributed to the creation of a river greatly less polluted in 1900 than it had been during the years of crisis between 1830 and 1860. Notions of what constituted impermissible levels of river contamination and how, scientifically, environmental deterioration should be defined and rectified, had nevertheless been transformed between the 1840s and the early twentieth century. Around the middle of the nineteenth century there had been no consensus about how the problem of pollution should be identified: all that was agreed was that to drink dirty water was generally inimical to good health and that clean rivers and human waste each had their rightful and ordered place in nature. The slowly widening appeal during the 1850s and 1860s of the germ or 'poison' theory of disease legitimated commitment to the idea that river pollution was both unhealthy and environmentally and economically insupportable. But innovation at the level of scientific theory could not provide a panacea to massive epidemiological danger, nor did it undermine residual commitment to the policy of removing waste from rivers and transporting it back to the land. The agricultural utilisation of sewage retained minority support into the 1890s—the notion that anything need be cast away, without use or profit, was anathema to the late Victorian ruling élite.

Even when it had been agreed that the water 'poison' and the pollution by which it was engendered were profoundly detrimental to health and must be eradicated, scientists had great difficulty in agreeing on ways and means. Water analysts searched for techniques which might reveal an organic proxy for the invisible 'disease matter', while medical men and social statisticians pragmatically insisted that, in the existing state of knowledge, the most advanced water purification techniques should be applied to every river suspected of being a threat to health. It was briefly believed, during the 1860s, that the 'poison' was as likely to yield up its mysteries to chemical as to biological speculation and investigation. Finally, however, 'proto-bacteriology' gave way not to a 'chemistry of disease' but to bacteriology proper; and chemical analysis, which had endured severe scientific and professional crises during the previous 20 years, entered a final, though complex, period of decline.

In terms of individuals, the debate about the Thames between the mid-1860s and the 1890s was dominated by Edward Frankland. A pessimist, he was influenced both by his knowledge of the large-scale water-transmitted epidemics of the 1840s and 1850s and, at first hand, by the destruction of life in the East End in 1866. Even when, in the 1890s, he had been converted to the fundamental tenets of bacteriology, Frankland was still periodically drawn back to his earlier preoccupation with 'previous sewage contamination'—the conviction that water once polluted with human faeces would carry the 'mark' of that contamination indefinitely and thus constitute a permanent threat to public health. And yet, paradoxically, no scientist was more deeply committed than Frankland to innovation in water purification processes. Nor—in the wider context—was any contemporary better able to bridge the gap between the 'social' and the 'scientific', or more ready to give full and scrupulous accounts of positions with which he disagreed. He commands respect both as a scientist and as a man.

The Thames had been central to society and economy in early-nineteenth-century London. A wretchedly poor riverside community scavenged a kind of existence from its mud flats; canny bargees surreptitiously dumped every kind of rubbish overboard when the police were overstretched; fishermen still netted an occasional salmon and hoped for a full revival of their trade. By the end of the century the river had become inaccessible to marginal groups and occupations. Formally embanked and hemmed in by bye-laws, it was now closed territory to all except those who had a legal and bureaucratic right to work on or near it. A formidable accumulation of official and scientific literature—proto-environmental and socio-medical surveys, engineering reports and geological research—had transformed the river into an object of study and simultaneously distanced it from everyday structures and processes. The Thames was no longer feared either as a bearer in its 'atmosphere' of 'plague-like' disease or as a symbol of urban disorder. But neither was it any longer fully integrated into the totality of material life.

It would be perverse to conclude a book of this type without making some general comments on the problem of pollution in the late twentieth century. That access to clean and safe water in Afro-Asia is still denied to entire communities and that very large numbers of people are therefore regularly struck down by cholera, typhoid, dysentery and the diarrhoeal infections, is too well known to need further emphasis. It has sometimes been argued that the position of the developing world is comparable with that of Western cultures afflicted by water-transmitted and water-related diseases during the nineteenth century. But such analogies lack historical validity—the plight of Afro-Asia is in this respect more extreme than anything experienced in Europe a century or more ago.

The determinants of this catastrophic situation are well documented. The colonial powers failed to invest adequately in essential public services. In the post-colonial era aid, earmarked for the improvement of water supplies, has been meagre and misapplied. And ecological and climatic change, particularly in Africa, now threaten future resources. In terms of methods of delivery in remote rural areas, little attention has been given until recently to the geographical and cultural idiosyncracies of specific regions, or to fundamental beliefs, of the kind described in earlier pages, about interactions between purity, pollution and the social order. (In an increasing number of development projects, however, anthropological insights are being deployed in order to avoid the most serious of earlier errors.)

In the West, meanwhile, environmental dilemmas and the sciences directed towards their resolution have changed and proliferated. One may usefully make mention here of atmospheric pollution, and, more specifically of lead emission from petrol: the brutal erosion of privacy and peace of mind by cars and planes: and visual and aesthetic pollution—people having no option but to wake up, day in, day out, and live with inescapable perceptions of drabness and inner-city decay. The reason why so few of these dilemmas have yet begun to be resolved has less to do with their 'complexity' than with constraints of social structure and politics. Thus the failure to legislate effectively against intolerably high levels of lead emission, now known to precipitate irreversible brain damage among infants and children, can be put down to an unwillingness to alienate massively powerful multinational corporations. But action has also been inhibited by cultural ambivalence towards mental retardation itself and an absence of participatory and non-stigmatising forums in which this ambivalence might be better understood and reduced.

In explicitly political terms, the undermining in Britain, in particular, of municipal and metropolitan autonomy and progressivism under a dominant ideology committed to privatisation and the neutralisation of collectivism has reduced investment in expertise required to deal with a wide range of environmental troubles. At an extreme this is an ideology which implies nothing less than a writing-off of large sections of cities as places in which lives can be meaningfully lived and the urban context improved or preserved. Such cynical pessimism does not, of course, go unchallenged. But it is becoming increasingly deeply embedded in every Western society—whether politically conservative or social democratic—dominated and preoccupied by seemingly permanent economic stagnation or recession. It is this tendency to disown the city, or, at best, to seek to contain its more disturbing oppositional manifestations, which starkly demonstrates the historical divide between late-nineteenth- and late-twentieth-century attitudes towards the urban condition. For, once they had achieved partial comprehension of the potential disruptiveness of the earlier Victorian city, a majority of medical men and scientists concerned with what is now called 'environmentalism' tentatively foresaw a period of orderly prosperity and 'improvement', disciplined enlightenment and control. To them, at least, whatever may have been their doubts about the explicitly political implications of municipalism, interventionism, and collectivism, 'the salvation of the city' was nothing less than a binding moral duty.

# Biographical Appendix

**William Budd** (1811–80) was educated in London, Edinburgh and Paris. He settled in Bristol in 1842 and was appointed physician at the Royal Infirmary. He undertook important work into the epidemiology of typhoid but also contributed to an understanding of tuberculosis, cancer and the diseases of animals. He became an FRS in 1871. (*Dictionary of National Biography* and Margaret Pelling, *Cholera, Fever and English Medicine 1825–1865* (Oxford, 1978))

**Sir William Crookes** (1832–1919) was a versatile Victorian man of science, more admired for his experimental ability than theoretical insight. He founded *Chemical News* in 1859 and in 1873 discovered the atomic weight of thallium. In later life Crookes specialised in spectroscopy, and aroused controversy as a result of involvement in psychical research. (*DNB*)

**William Farr** (1807–83) came from a humble background. He studied in Paris and London and in 1838 became compiler of abstracts at the Registrar-General's Office. For the next 40 years, Farr was a highly influential, though at times unpredictable, figure in the fields of epidemiology and social statistics. (J M Eyler, *Victorian Social Medicine: The Ideas and Methods of William Farr* (Baltimore, 1979))

**Sir Edward Frankland** (1825–99) was the 'founding father' of valency theory and also undertook seminal investigations into the causes of and remedies for river pollution. He worked with Liebig in Germany before being appointed at a very young age to a professorship at the newly established Owens College, Manchester. He became an FRS in 1853 and was professor of chemistry at the Royal Institution between 1863 and 1868 and at the Royal College of Chemistry (later united with the Royal College of Mines) from 1865 to 1885. His laboratory skills were greatly admired by contemporaries. (*DNB*)

**Edward Headlam Greenhow** (1814–88) studied medicine at Edinburgh and Montpellier, France. In 1855 he was appointed lecturer in public health at St Thomas's Hospital. Moving to the Middlesex Hospital in 1861 he became a lecturer in medical jurisprudence. He undertook significant surveys into the epidemiology of diarrhoea, diphtheria and bronchitis and also had a long-standing interest in Addison's disease. (*DNB* and Royston Lambert, *Sir John Simon and English Social Administration 1816–1904* (1963))

**Henry Letheby** (1816–76) was lecturer in chemistry at the London Hospital. He succeeded John Simon as medical officer to the City of London in 1855. He was in constant demand as an analytical chemist and also worked for the government as a metropolitan gas inspector. Letheby wrote extensively on the chemical composition of food. (*DNB* and Lambert *op. cit.*)

**John Martin** (1789–1854), whose stock has risen in recent years, was the son of a fencing master. He improved his artistic technique by painting on china and glass and then went on to produce a series of canvasses devoted to 'sublime' and 'catastrophic' subjects such as 'The Fall of Babylon' and 'The Destruction of Sodom and Gomorrah'. But he soon fell out with the artistic establishment and lost money on numerous commercial ventures designed to save London from environmental destruction. He is alleged to have died from under-nourishment, convinced that food would worsen rather than improve his health while recovering from a 'paralytic' illness. (*DNB* and M L Pendered, *John Martin, Painter* (1923))

**Sir Arthur Newsholme** (1857–1943) was an influential early-twentieth-century expert on public health and public policy. He entered general practice in Clapham before being appointed medical officer of health at Brighton in 1888. In the following year he published his authoritative *The Elements of Vital Statistics* and in 1908 became Medical Officer to the Local Government Board, a position he held until the establishment of the Ministry of Health in 1919. During retirement Newsholme travelled widely and wrote, among much else, *Fifty Years in Public Health* (1935) and *The Last Thirty Years in Public Health* (1936). (*DNB*)

**William Miller Ord** (1834–1902) was a life-long personal friend and colleague of John Simon and combined physiological research and teaching at St Thomas's with a number of revealing epidemiological surveys. He played an important role in the elucidation of myxoedema. (*DNB* and Lambert *op. cit.*)

**John Netten Radcliffe** (1826–84) was educated in Leeds and served as a surgeon in the Crimea. At the end of the Crimean War he was appointed medical superintendent of the Hospital for the Paralysed and Epileptic in Queen Square, London. He made a national reputation for himself with his investigation of the cholera epidemic of 1866 and in 1871 took up the position of assistant medical officer at the Local Government Board. He possessed a fine analytic intelligence. (*DNB*)

**Sir John Simon** (1816–1904) retained a life-long commitment to pathology, his first field of interest, and was lecturer in that subject at St Thomas's from 1847 to 1870. But he came to national prominence in the sphere of public health as Medical Officer of Health to the Privy Council. Between 1858 and the early 1870s, Simon commissioned a revolutionary series of surveys into national and regional patterns of disease and neglected aspects of scientific medicine. With the formation of the Local Government Board in 1871 Simon's fortunes went into decline. A complex man, he was equally at home in artistic and scientific culture. (Lambert *op. cit.*)

**John Snow** (1813–58) was born in York and studied medicine in London. He became a consultant anaesthetist and attended the birth of Princess Beatrice in 1857. But it is as an epidemiologist and interpreter of the interactions between dirty water and cholera that Snow is now remembered. By the standards of the 1850s, his work displayed a rigorous empiricism and a refusal to appeal to what he considered to be inherently unlikely residual explanations. Snow was a teetotaler and a vegetarian. (*DNB*)

**Charles Meymott Tidy** (1843–92) studied under Henry Letheby at the London Hospital and then, following 10 years in general practice, succeeded him in 1876 as professor of chemistry, medical jurisprudence and public health. He published extensively on the treatment of sewage and toxicology. (*DNB*)

**James Alfred Wanklyn** (1834–1906) served his apprenticeship with Frankland, Bunsen and Lyon Playfair and held the chair of chemistry at the London Institution between 1863 and 1870. For the bulk of his career, however, he worked in a non-academic context and acted, at various times, as public analyst for Buckingham, High Wycombe, Peterborough and Shrewsbury. In the early 1870s Wanklyn devoted himself to an examination of the quality of milk consumed in London but, in this field, as in water analysis, he became involved in lengthy and often acrimonious debate. (*DNB*)

# Bibliography

## Unpublished Sources

At the Wellcome Unit for the History of Medicine, Oxford:
  *Minutes of the Metropolitan Association of Medical Officers of Health.*
At the Thames Conservancy Archive, Wapping:
  *Minutes and Committee Proceedings of the Thames Conservancy Board.*
At the Greater London Council History Library:
  *Reports of the Statistical Committee of the Metropolitan Asylums Board.*
At the Public Record Office, Kew:
  *Cabinet Papers* on environmental legislation.

## Official and Parliamentary Publications

Annual *Reports* of the Metropolitan Medical Officers of Health.
Annual *Reports* of the Medical Officer of Health to the London County Council.
Annual *Reports* and *Supplements* of the Registrar-General.
Annual *Reports* of the Metropolitan Board of Works.
Annual *Reports* of the Thames Conservators.
Annual *Reports* of the Local Government Board.
Hansard.
*Royal Commission on the Water Supply of the Metropolis* PP 1828:IX:47.
*Select Committee on the Sewers of the Metropolis* PP 1834:XV:197.
*Select Committee on the Health of Towns* PP 1840:XI:277.
*Select Committee on the Thames Embankment* PP 1840:XII:271.
*Select Committee (HL) on the Supply of Water to the Metropolis* PP 1840:XII:159.
*Royal Commission on the Improvement of the Metropolis* PP 1844:XV:1.
*Select Committee on Metropolitan Sewage Manure* PP 1846:X:535.
*Metropolitan Sanitary Commission: First Report* PP 1847–8:XXXII:253.
*Report of the General Board of Health on the Epidemic Cholera of 1848 and 1849* PP 1850:XXI:3.
*Report of the General Board of Health on the Supply of Water to the Metropolis* PP 1850:XXII:1.

*Select Committee on the Metropolis Water Bill* PP 1851:XV:1.

*Report on the Chemical Quality of the Supply of Water to the Metropolis* PP 1851: XXIII:401.

*Report on the Mortality of Cholera in England 1848–9* (HMSO, 1852).

*Select Committee on Public Health and Nuisances Removal Bill* PP 1854–5:XIII:413.

*Report of the Committee for Scientific Enquiries in Relation to the Cholera Epidemic of 1854* PP 1854–5:XXI:135.

*Report of the Medical Council in relation to the Cholera Epidemic of 1854* PP 1854–5:XLV:1.

*Report on the Chemical Quality of the Supply of Water to the Metropolis* PP 1856:LII: 255.

*Report on the Work of the Metropolitan Water Companies* PP 1856:LII:275.

*Report on the Last Two Cholera Epidemics of London as Affected by the Consumption of Impure Water* PP 1856:LII:257.

'Proceedings in Reference to the London Sewage Nuisance' (William Ord) *Second Report of the Medical Officer of the Privy Council* PP 1860:XXIX:255.

'Proceedings in Reference to the Diarrhoeal Districts of England' (E H Greenhow) *Second Report of the Medical Officer of the Privy Council* PP 1860:XXIX:259.

*Select Committee on the Thames Conservancy Bill* PP 1863:XII:1.

*Select Committee on the Sewage of the Metropolis* PP 1864:XIV:1.

'Letter to Treasury . . . on pollution of streams' PP 1864:L:327.

*Select Committee on the Thames Navigation Bill* PP 1866:XII:491.

*Rivers Pollution Commission: First Report: River Thames* PP 1866:XXXIII:1.

'Copy of a Letter addressed by direction of the Secretary of State for the Home Department to the Metropolitan Water Companies. . . .' PP 1866:LXVI: 671.

*Select Committee on the East London Water Bills* PP 1867:IX:1.

*Rivers Pollution Commission: Second Report: River Lea* PP 1867:XXXIII:29.

'Mr J. Netten Radcliffe on Cholera in London, and especially in the Eastern Districts' *Ninth Report of the Medical Officer of the Privy Council* PP 1867: XXXVII:264.

'Report of Captain Tyler to the Board of Trade in Regard to the East London Waterworks Company' PP 1867:LVIII:441.

*Report on the Cholera Epidemic of 1866 in England* PP 1867–8:XXXVII:1.

*Royal Commission on Water Supply* PP 1868–9:XXXIII:1.

'Report by J. Netten Radcliffe on the Turbidity of Water Supplied by Certain London Companies' *Twelfth Annual Report of the Medical Officer of the Privy Council* PP 1870:XXXVIII:767.

*Select Committee on Metropolis (No 2) Water Bill* PP 1871:XI:1.

*Metropolitan Sanitary and Street Improvements* PP 1872:XLIX:585.

*Select Committee (HL) Pollution of Rivers Bill* PP 1873:IX:5.

'Report on an Outbreak of Enteric Fever in Marylebone and the Adjoining Parts of London by J. Netten Radcliffe and W. H. Power' *Supplementary Report of the Medical Officer of the Privy Council and Local Government Board* PP 1874:XXXI:137.

*Royal Commission on the Sanitary Laws: Second Report* PP 1874:XXXI:603.

*Rivers Pollution Commission: Sixth Report: Domestic Water Supply of Great Britain* PP 1874:XXXIII:311.

'Paper by the Rt Hon Lord Robert Montagu MP on Watershed Boards. . . .'
PP 1876:LX:609.
*Select Committee on London Water Supply* PP 1880:X:111.
*Royal Commission on Metropolitan Sewage Discharge* PP 1884:XLI:1 and
PP 1884–5:XXXI:341.
*Select Committee on Rivers Pollution (Lee)* PP 1886:XI:267.
*Supplement in Continuation of the Report of the Medical Officer of Health for 1887* PP
1889:XXXV:5.
*Royal Commission on the Water Supply of the Metropolis* PP 1893–4:XL(Pts I and
II):1.
*Royal Commission on Water Supply within the Limits of the Metropolitan Water
Companies* PP 1900:XXXVIII (Pt I):1.

## Contemporary Books and Articles

Barnes R 1858 'Is the Thames Pernicious?', *Journal of Public Health and Sanitary
Review* **iv**.
Bazalgette J W 1864–5 'On the Main Drainage of London and the Interception
of Sewage from the River Thames', *Minutes and Proceedings of the Institute of
Civil Engineers* **xxiv**.
Bolton F and Scratchley P A 1888 *The London Water Supply* second edition.
Buck W E 1885–6 'On Infantile Diarrhoea', *Transactions of the Sanitary Institute
of Great Britain* **vii**.
Budd W 1873 *On Typhoid Fever*.
Cassall C E and Whitelegge B A 1883–4 'Remarks on the Examination of
Water for Sanitary Purposes', *Transactions of the Society of Medical Officers of
Health*.
Childe-Pemberton W S 1909 *Life of Lord Norton*.
Childs C 1898 'Water-borne Typhoid Fever', *Journal of the Sanitary Institute of
Great Britain* **xix**.
Davies S 1908 'Twenty Years' Metropolitan Advance in Preventive Medicine',
*Public Health* **xxi**.
Dudfield R 1906 'History of the Society of Medical Officers of Health', *Public
Health* Jubilee Number.
Fishbourne E H 1882 *The Thames Conservancy*.
Fletcher J 1845 'Historical and Statistical Account of the Present System of
Supplying the Metropolis with Water', *Journal of the Statistical Society* **viii**.
Flower L 1887–8 'On the Fouling of Streams', *Transactions of the Sanitary
Institute of Great Britain* **ix**.
—— 1876 'Sewage Treatment: More Especially as Affecting the Pollution of
the River Lee', *Journal of the Royal Society of Arts* **xxiv**.
Frankland E 1876 'On Some Points in the Analysis of Potable Water', *Journal
of the Chemical Society* **xxviii**.
Frankland P 1886–7 'On the Filtration of Water for Town Supply', *Transactions
of the Sanitary Institute of Great Britain* **viii**.
Guy W 1855 'On the Fluctuations in the Number of Births, Deaths and
Marriages in the Metropolis . . . 1840–54', *Journal of the Statistical Society*
**xviii**.

Harrison J T 1853–4 'On the Drainage of the District South of the Thames', *Minutes and Proceedings of the Institute of Civil Engineers* **xiii**.

Hassall A H 1876 *Food: Its Adulterations and the Methods for their Detection*.

Jephson A H 1907 *The Sanitary Evolution of London*.

Jones H R 1894 'The Perils and Protection of Infant Life', *Journal of the Statistical Society* **lvii**.

Leigh A G 1903 'Manufacturers and the Rivers Pollution Prevention Act', *Journal of the Sanitary Institute of Great Britain* **xxiv**.

Longstaff G B 1884 'The Recent Decline in the English Death Rate considered in connection with the Causes of Death', *Journal of the Statistical Society* **xlvii**.

—— 1884–5 'The Seasonal Prevalence of Continued Fever in London', *Transactions of the Epidemiological Society*.

Miller J W 1909–10 'Filtration and Other Methods of Purification, on a Large Scale, of River Water used for Drinking Purposes', *Public Health* **xxiii**.

Milnes-Gaskell C 1903 'On the Pollution of our Rivers', *Nineteenth Century* **xlvi** (July).

Murchison C 1884 *A Treatise on the Continued Fevers of Great Britain* third edition (edited by W Cayley).

Murphy S F 1879–80 'Autumnal Diarrhoea', *Transactions of the Society of Medical Officers of Health*.

Newsholme A 1896 'The Spread of Enteric Fever by means of Sewage-Contaminated Shellfish', *Journal of the Sanitary Institute of Great Britain* **xvii**.

—— 1899 'A Contribution to the Study of Epidemic Diarrhoea', *Public Health* **xii**.

Nicholas G E 1857 'On the Drainage of the Metropolis by a Continuous Water Flow', *Journal of Public Health and Sanitary Review* **iii**.

Orton T 1866 'Special Report by Thomas Orton, MOH, Limehouse, on the Cholera Epidemic of 1866'.

Palmberg A 1893 *A Treatise on Public Health and its Applications* (translated by A Newsholme).

Parkes L 1887–8 'On Water Analysis', *Transactions of the Sanitary Institute of Great Britain* **ix**.

—— 1892 'The Air and Water of London: Are They Deteriorating?', *Transactions of the Sanitary Institute of Great Britain* **xiii**.

Radcliffe J N 1863 'Reports on Epidemics', *Transactions of the Epidemiological Society* **ii**.

Ramsay W 1908 'Rivers Pollution', *Journal of the Royal Sanitary Institute* **xxix**.

Richards H E and Payne W H C 1899 *London Water Supply* second edition (edited by J P H Soper).

Robinson H 1855–6 'On the Past and Present Condition of the River Thames', *Minutes and Proceedings of the Institute of Civil Engineers* **xv**.

Russell H W 1901 'Prevention of Pollution within the Thames Watershed', *Journal of the Sanitary Institute of Great Britain* **xxii**.

Saunders C E 1886–7 'Legislation for the Purification of Rivers and its Failures', *Transactions of the Society of Medical Officers of Health*.

Scott-Moncrieff W D 1909 'River Pollution: Its Ethics, Aesthetics and Hygiene', *Journal of the Royal Sanitary Institute* **xxx**.

Shadwell A 1899 *The London Water Supply*.

Smith P C 1906–7 'The Metropolitan Main Drainage and its Effects on Mortality 1857–1907', *Public Health* **xix**.

Snow J 1936 *On Cholera* edited by Wade Hampton Frost, New York.

Spence F 1893 'How to Stop River Pollution', *Contemporary Review* (September).

Symons G J 1880 'Health and Pure Water', *Transactions of the Sanitary Institute of Great Britain* **i**.

Thomas A E 1908 'River Water Supplies and Filtered Waters', *Public Health* **xxi**.

Thorne R T 1888 *The Progress of Preventive Medicine during the Victorian Era*.

Ward F O 1858 'Purification of the Thames' *Journal of Public Health and Sanitary Review* **iv**.

Wilson G 1895 'River-water and Dissemination of Disease', *Journal of State Medicine* **iii**.

## Secondary Books and Articles

Ackerknecht E H 1948 'Anticontagionism between 1821 and 1867', *Bulletin of the History of Medicine* **22** 562–93.

Alexander M 1971 *Microbial Ecology*, New York.

American Water Works Association 1950 *Water Quality and Treatment* second edition, New York.

Anderson J R L 1970 *The Upper Thames*.

Appleby A 1980 'The Disappearance of Plague: A Continuing Puzzle', *Economic History Review* **33** 161–73.

Armstrong A 1974 *Stability and Change in an English Town: A Social Study of York 1801–1851*, Cambridge.

Ayers G M 1971 *England's First State Hospitals and the Metropolitan Asylums Board, 1867–1930*.

Balston T 1947 *John Martin, 1789–1854: His Life and Works*.

Banks H S 1949 *The Common Infectious Diseases*.

Barnes S B 1974 *Scientific Knowledge and Sociological Theory*.

Berg A 1973 *The Nutrition Factor: Its Role in National Development*, Washington, DC.

Best G 1971 *Mid-Victorian Britain 1850–1875*.

Blake R 1966 *Disraeli*.

Bloor D 1976 *Knowledge and Social Imagery*.

Braudel F 1972 *The Mediterranean and the Mediterranean World in the Age of Philip II* volumes I and II, translated by S Reynolds.

—— 1973 *Capitalism and Material Life 1400–1800*, translated by M Kochan.

—— 1977 *Afterthoughts on Material Civilization and Capitalism*, translated by P M Ranum, Baltimore.

Briggs A 1960–1 'Cholera and Society in the Nineteenth Century', *Past and Present* **19** 79–96.

Bulloch W 1938 *The History of Bacteriology*, Oxford.

Cannadine D and Reeder D (eds) 1982 *Exploring the Urban Past: Essays in Urban History*, Cambridge.

Cassedy J H 1962 *Charles V Chapin and the Public Health Movement*, Cambridge, MA.

Chalmers A K 1930 *The Health of Glasgow: An Outline*, Glasgow.

Chaloner W H 1950 *The Social and Economic Development of Crewe*.

Church R A 1975 *The Great Victorian Boom*.

Cobb R 1975 'The Seine' in *Paris and Its Provinces 1792–1802*, Oxford, 57–86.

Cooter R J 1982 'Anticontagionism and History's Medical Record' in P Wright and A Treacher (eds), *The Problem of Medical Knowledge*, Edinburgh, 87–108.

—— 1984 *The Cultural Meaning of Popular Science: Phrenology and the Organization of Consent in Nineteenth Century Britain*, Cambridge.

Creighton C 1965 *A History of Epidemics in Britain*, two volumes, revised edition.

Crellin J K 1968 'The Dawn of the Germ Theory: Particles, Infection and Biology' in F N L Poynter (ed), *Medicine and Science in the 1860s*, 57–76.

Cullen M J 1975 *The Statistical Movement in Early Victorian Britain*, Brighton.

Davin A 1978 'Imperialism and Motherhood', *History Workshop Journal* **5** 9–67.

De S N 1961 *Cholera: Its Pathology and Pathogenesis*, Edinburgh.

Dickinson H W 1954 *The Water Supply of Greater London*.

Douglas M 1966 *Purity and Danger: An Analysis of Concepts of Pollution and Taboo*.

—— 1975 *Implicit Meanings: Essays in Anthropology*.

Douglas M and Wildavsky A 1982 *Risk and Culture: An Essay on the Selection of Technological and Environmental Dangers*, Berkeley.

Durey M 1974 *The First Spasmodic Cholera Epidemic in York, 1832*. York.

—— 1979 *The Return of the Plague: British Society and the Cholera 1831–2*, Dublin.

Dyhouse C 1978–9 'Working Class Mothers and Infant Mortality in England 1895–1914', *Journal of Social History* **12** 248–67.

Dyos H J 1961 *Victorian Suburb: A Study in the Growth of Camberwell*, Leicester.

Erlich P R and Erlich A H 1970 *Population, Resources, Environment: Issues in Human Ecology*, San Francisco.

Eyler J M 1973 'William Farr on the Cholera: The Sanitarian's Disease Theory and the Statistician's Method', *Journal of the History of Medicine* **28** 79–100.

—— 1979 *Victorian Social Medicine: The Ideas and Methods of William Farr*, Baltimore.

—— 1980 'The Conversion of Angus Smith: The Changing Role of Chemistry and Biology in Sanitary Science, 1850–1880', *Bulletin of the History of Medicine* **54** 216–34.

Figlio K 1979 'Sinister Medicine? A Critique of Left Approaches to Medicine', *Radical Science Journal* **9** 14–68.

Finer S E 1952 *The Life and Times of Sir Edwin Chadwick*.

Foucault M 1970 *The Order of Things: The Archaeology of the Human Sciences*.

Frazer W M 1950 *A History of English Public Health 1834–1939*.

Gale A H 1945 'A Century of Changes in the Mortality and Incidence of the Principal Infections of Childhood', *Archives of Diseases of Childhood* **20** 2–21.

—— 1959 *Epidemic Diseases*.

Gash N 1979 *Aristocracy and People: Britain 1815–1865*.

Gladstone G P 1962 'Pathogenicity and Virulence of Microorganisms' in H Florey (ed), *General Pathology*, third edition, 692–722.

Goodall E W 1934 *A Short History of the Epidemic Infectious Diseases.*
Greenwood M 1935 *Epidemics and Crowd Diseases.*
Hall P G 1962 *The Industries of London since 1861.*
Hamlin C 1982 'Edward Frankland's Early Career as London's Official Water Analyst 1865–1876: The Context of "Previous Sewage Contamination"', *Bulletin of the History of Medicine* **56** 56–76.
Hammond J L 1930 'The Industrial Revolution and Discontent', *Economic History Review* **11** 215–28.
Hardy A 1984 'Water and the Search for Public Health in London in the Eighteenth and Nineteenth Centuries', *Medical History* **28** 250–82.
Hartwell R M 1971 *The Industrial Revolution and Economic Growth.*
Hennock E P 1973 *Fit and Proper Persons: Ideal and Reality in Nineteenth Century Urban Government.*
Hobsbawm E J 1964 *Labouring Men.*
Howard-Jones N 1972 'Cholera Therapy in the Nineteenth Century', *Journal of the History of Medicine* **27** 373–95.
Ignatieff M 1978 *A Just Measure of Pain: The Penitentiary in the Industrial Revolution 1750–1850.*
Jensen J V 1970–1 'The X Club: Fraternity of Victorian Scientists', *British Journal of the History of Science* **5** 63–72.
Jordanova L J and Porter R S (eds) 1979 *Images of the Earth: Essays in the History of the Environmental Sciences.*
Khin-Maung U *et al* 1985 'Effect on Clinical Outcome of Breast Feeding during Acute Diarrhoea', *British Medical Journal* **290** 587–9.
Kitson Clark G 1967 *An Expanding Society: Britain 1830–1900*, Cambridge.
Klingender F D 1947 *Art and the Industrial Revolution.*
Kuhn T S 1962 *The Structure of Scientific Revolutions*, Chicago.
Ladurie E Le Roy 1972 *Times of Feast, Times of Famine: A History of Climate since the Year 1000*, translated by B Bray.
—— 1979 *The Territory of the Historian*, translated by S and B Reynolds.
—— 1981 *The Mind and Method of the Historian*, translated by S and B Reynolds.
Lambert R 1963 *Sir John Simon, 1816–1904, and English Social Administration.*
Lewis R A 1952 *Edwin Chadwick and the Public Health Movement 1832–1854.*
Lipschutz D 1968 'The Water Question in London', *Bulletin of the History of Medicine* **42** 510–25.
MacDonagh O 1961 *A Pattern of Government Growth: The Passenger Acts and their Enforcement 1800–1860.*
Macfarlane A 1978 *The Origins of English Individualism*, Oxford.
McKeown T 1976 *The Modern Rise of Population.*
McKeown T and Record R G 1962 'Reasons for the Decline of Mortality in England and Wales during the Nineteenth Century', *Population Studies* **16** 94–122.
MacLaren A A 1976 'Bourgeois Ideology and Victorian Philanthropy: The Contradictions of Cholera' in A A MacLaren (ed), *Social Class in Scotland: Past and Present*, Edinburgh, 36–54.
MacLeod R M 1965 'The Alkali Acts Administration 1863–84: The Emergence of the Civil Scientist', *Victorian Studies* **9** 85–112.

—— 1968 'Government and Resource Conservation: The Salmon Acts Administration', *Journal of British Studies* **7** 144–50.

—— 1970 'The X-Club: A Social Network of Science in Late Victorian England', *Royal Society: Notes and Records* **24** 305–22.

McNeill W H 1977 *Plagues and Peoples*, Oxford.

Morley D 1973 *Paediatric Priorities in the Developing World.*

Morris R J 1976 *Cholera 1832: The Social Response to an Epidemic.*

Mulkay M 1979 *Science and the Sociology of Knowledge.*

Newsholme A 1923 *The Elements of Vital Statistics.*

—— 1935 *Fifty Years in Public Health.*

Owen D 1982 *The Government of Victorian London 1855–1889: The Metropolitan Board of Works, the Vestries and the City Corporation*, edited by Roy MacLeod, Cambridge, MA.

Parry-Jones W L 1972 *The Trade in Lunacy: A Study of Private Madhouses in England in the Eighteenth and Nineteenth Centuries.*

Parsons L and Barling S (eds) 1954 *Diseases of Infancy and Childhood*, vol I, Oxford.

Partington J R 1964 *A History of Chemistry*, vol 4.

Passmore J 1974 *Man's Responsibility for Nature.*

Patrick A 1955 *The Enteric Fevers*, Edinburgh.

Paul H 1964 *The Control of Diseases Social and Communicable*, second edition, Edinburgh.

Pelling M 1978 *Cholera, Fever and English Medicine 1825–1865*, Oxford.

Pendered M L 1923 *John Martin, Painter.*

Phelps Brown E H 1960 *The Growth of British Industrial Relations.*

Porter R 1982 *English Society in the Eighteenth Century.*

Razzell P 1977 *The Conquest of Smallpox*, Firle.

Ritchie J 1937 'Enteric Fever', *British Medical Journal* ii 160–3.

Roberts R 1971 *The Classic Slum: Salford Life in the First Quarter of the Century*, Manchester.

Robson W A 1939 *The Government and Misgovernment of London.*

Rosen G 1958 *A History of Public Health*, New York.

—— 1973 'Disease, Debility and Death' in H J Dyos and M Wolff (eds), *The Victorian City: Images and Reality*, vol II 625–67.

Rosenberg C 1962 *The Cholera Years: The United States in 1832, 1849 and 1866*, Chicago.

Sheppard F 1971 *London 1808–1870: The Infernal Wen.*

Shrewsbury J F D 1971 *A History of Bubonic Plague in the British Isles*, Cambridge.

Singer C and Underwood E A 1962 *A Short History of Medicine*, second edition.

Skinner Q 1969 'Meaning and Understanding in the History of Ideas', *History and Theory* **8** 3–53.

Slack P 1981 'The Disappearance of Plague: An Alternative View', *Economic History Review* **34** 469–76.

Smith F B 1979 *The People's Health 1830–1910.*

Smith P 1967 *Disraelian Conservatism and Social Reform.*

Spink W 1978 *Infectious Diseases: Prevention and Treatment in the Nineteenth and Twentieth Centuries*, Folkestone.

Stedman Jones G 1971 *Outcast London: A Study in the Relationship between Classes in Victorian Society*, Oxford.

Stewart F S 1962 *Bigger's Handbook of Bacteriology*, eighth edition.

Stone L 1965 *The Crisis of the Aristocracy*, Oxford.

Taylor A J 1972 *Laissez-faire and State Intervention in Nineteenth Century Britain*.

—— (ed) 1975 *The Standard of Living in Britain in the Industrial Revolution*.

Taylor J 1970 'Infectious Infantile Entiritis, Yesterday and Today', *Proceedings of the Royal Society of Medicine* **58** 1294–301.

Thacker F S 1968 *The Thames Highway*, 2 vols.

Thomas K 1983 *Man and the Natural World: Changing Attitudes in England 1500–1800*.

Thompson E P 1975 *Whigs and Hunters*.

Thresh J C and Beale J F 1925 *The Examination of Waters and Water Supplies*, third edition.

von Tunzelmann G N 1979 'Trends in Real Wages 1750–1850, Revisited', *Economic History Review* **32** 33–49.

Vincent J R 1966 *The Formation of the Liberal Party 1857–1868*.

Walker G E 1957 *The Thames Conservancy 1857–1957*.

Whetham E H 1965–6 'The London Milk Trade 1860–1900', *Economic History Review* **17** 369–80.

Williams R 1973 *The Country and the City*.

Wilson G S and Miles A A (eds) 1964 *Topley and Wilson's Principles of Medicine and Immunity*, fifth edition.

Wohl A S 1977 *The Eternal Slum: Housing and Social Policy in Victorian London*.

—— 1983 *Endangered Lives: Public Health in Victorian Britain*.

Wood L B 1982 *The Restoration of the Tidal Thames*, Bristol.

Woods R and Woodward J (eds) 1984 *Urban Disease and Mortality in Nineteenth Century England*.

Wrigley E A 1967 'A Simple Model of London's Importance in Changing English Society and Economy', *Past and Present* **37** 44–70.

Wrigley E A and Schofield R S 1981 *The Population History of England 1541–1871: A Reconstruction*.

Young R M 1973 'The Historiographic and Ideological Contexts of the Nineteenth Century Debate on Man's Place in Nature' in M Teich and R M Young (eds), *Changing Perspectives in the History of Science*, 344–438.

—— 1977 'Science *is* Social Relations', *Radical Science Journal* **5** 65–129.

## Theses

Lambert R J 1959 'State Activity in Public Health 1858–71', *PhD Thesis* (University of Cambridge).

MacLeod R M 1967 'Specialist Policy in Government Growth', *PhD Thesis* (University of Cambridge).

Mukhopadhyay A K 1971 'The Politics of London Water Supply, 1871–1971', *PhD Thesis* (University of London).

Richards T 1982 'River Pollution Control in Industrial Lancashire, 1848–1939', *PhD Thesis* (University of Lancaster).

# Index